COMPUTER SCIENCE, TECHNOLOGY AND APPLICATIONS

THE FUNDAMENTALS OF SEARCH ALGORITHMS

COMPUTER SCIENCE, TECHNOLOGY AND APPLICATIONS

Additional books and e-books in this series can be found on Nova's website under the Series tab.

COMPUTER SCIENCE, TECHNOLOGY AND APPLICATIONS

THE FUNDAMENTALS OF SEARCH ALGORITHMS

ROBERT A. BOHM
EDITOR

Copyright © 2021 by Nova Science Publishers, Inc.

All rights reserved. No part of this book may be reproduced, stored in a retrieval system or transmitted in any form or by any means: electronic, electrostatic, magnetic, tape, mechanical photocopying, recording or otherwise without the written permission of the Publisher.

We have partnered with Copyright Clearance Center to make it easy for you to obtain permissions to reuse content from this publication. Simply navigate to this publication's page on Nova's website and locate the "Get Permission" button below the title description. This button is linked directly to the title's permission page on copyright.com. Alternatively, you can visit copyright.com and search by title, ISBN, or ISSN.

For further questions about using the service on copyright.com, please contact:
Copyright Clearance Center
Phone: +1-(978) 750-8400   Fax: +1-(978) 750-4470   E-mail: info@copyright.com.

NOTICE TO THE READER

The Publisher has taken reasonable care in the preparation of this book, but makes no expressed or implied warranty of any kind and assumes no responsibility for any errors or omissions. No liability is assumed for incidental or consequential damages in connection with or arising out of information contained in this book. The Publisher shall not be liable for any special, consequential, or exemplary damages resulting, in whole or in part, from the readers' use of, or reliance upon, this material. Any parts of this book based on government reports are so indicated and copyright is claimed for those parts to the extent applicable to compilations of such works.

Independent verification should be sought for any data, advice or recommendations contained in this book. In addition, no responsibility is assumed by the Publisher for any injury and/or damage to persons or property arising from any methods, products, instructions, ideas or otherwise contained in this publication.

This publication is designed to provide accurate and authoritative information with regard to the subject matter covered herein. It is sold with the clear understanding that the Publisher is not engaged in rendering legal or any other professional services. If legal or any other expert assistance is required, the services of a competent person should be sought. FROM A DECLARATION OF PARTICIPANTS JOINTLY ADOPTED BY A COMMITTEE OF THE AMERICAN BAR ASSOCIATION AND A COMMITTEE OF PUBLISHERS.

Additional color graphics may be available in the e-book version of this book.

Library of Congress Cataloging-in-Publication Data

ISBN: 978-1-53619-007-6

Published by Nova Science Publishers, Inc. † New York

CONTENTS

Preface		vii
Chapter 1	The Fundamentals of Heuristic Local Search Algorithms for the Traveling Salesman Problem *Weiqi Li*	1
Chapter 2	Biometric Data Search Algorithm *Stella Metodieva Vetova*	45
Chapter 3	Differential Evolution for Solving Continuous Search Space Problems *Omar Andres Carmona Cortes and Hélder Pereira Borges*	75
Index		99

PREFACE

Heuristic local search algorithms are used to find "good" solutions to the NP-hard combinatorial optimization problems that cannot be solved using analytical methods. Chapter one discusses the characterization and computation of heuristic local search algorithm for the Traveling Salesman Problem (TSP) from the perspective of dynamical systems. The purpose of chapter 2 is to show the practical application of CBIR technology in the security and protection of personal data, access to classified documents and objects, identification of illegal attacks that are part of the social life of the present and future of mankind. Continuous search space problems are difficult problems to solve because the number of solutions is infinite. Moreover, the search space gets more complex as we add constraints to the problem. In this context, chapter 3 aims to show the usage of the differential evolution algorithm for solving continuous search space problems using unconstrained functions and a constrained real-world problem.

Chapter 1 - Heuristic local search algorithms are used to find "good" solutions to the NP-hard combinatorial optimization problems that cannot be solved using analytical methods. This chapter discusses the characterization and computation of heuristic local search algorithm for the Traveling Salesman Problem (TSP) from the perspective of dynamical systems. A heuristic local search system is essentially in the domain of dynamical systems. Like many other dynamical systems, a local search system has an attracting property that drives the search trajectories to converge to a small region, called a solution attractor, in the solution space. The study of the solution attractor in the solution space can provide the answer to the question: "Where do the search trajectories go in a heuristic local search system?" The solution attractor collects all locally optimal solutions around the globally optimal solutions. This chapter, using the TSP as the study problem, describes the behavior of search trajectories and properties of the solution attractor in a local search system. Based on these properties of the solution attractor, a novel global optimization algorithm – Attractor-Based Search System (ABSS) – is introduced. This algorithm provides the answer to the question: "How can we use efficient heuristic local search for global optimization?" The ABSS combines a local search process and an exhaustive search procedure. The local search process constructs the solution attractor of the local search system, and the exhaustive search procedure identifies the best solutions in the solution attractor. This chapter describes how the ABSS meets the requirements of global optimization system. The computational complexity of the ABSS for the TSP is also discussed.

Chapter 2 - The purpose of the outlined content is to show the practical application of CBIR technology in the security and protection of personal data, access to classified documents and objects, identification of illegal attacks that are part of the social life of the present and future of mankind. It involves comparison of the disadvantages and advantages of content-based image retrieval techniques with the use of local and global features in biometric data. The main idea is to offer the optimal application for security and protection. In pursuit of this goal, the following tasks are envisaged:

1. Design of algorithms with the use of local characteristics and
2. different similarity distance measures;
3. Design of algorithms with the use of global characteristics using
4. different similarity distance measures;
5. Conducting experimental studies on the algorithms of items 1 and 2;
6. Comparative analysis of the data obtained in item 3.

Considering this, two algorithms with different similarity distance measures were developed using the two techniques, and in this case Hausdorff distance and Euclidean distance were chosen as distance metrics. For the representation of a two-dimensional signal through its feature vectors, the decomposition of the signal to approximate and detailed coefficients by the 2D Dual - Tree Complex Wavelet Transform was used. The concept of this process is described visually by graphic illustrations, diagrams, images and analyzes. Also, a description of the research methodology, test databases and used software is included. To evaluate the effectiveness of the designed algorithms according to both techniques, test studies were conducted to evaluate the accuracy of the extracted result, the retrieval time, the number of retrieved images. The results are presented graphically, in tables and by image. At the end of the chapter, a comparative analysis between the algorithms is made on the base of the results obtained, which shows the advantages and disadvantages of both techniques and their application in the recognition of biometric data.

Chapter 3 - Continuous search space problems are difficult problems to solve because the number of solutions is infinite. Moreover, the search space gets more complex as we add constraints to the problem. In this context, this chapter aims to show the usage of the differential evolution algorithm for solving continuous search space problems using unconstrained functions and a constrained real-world problem. Six different mutation strategies were implemented: /DE/Rand/01, /DE/Best/01, /DE/Rand/02, /DE/Best/02, /DE/Rand-to-Best/01, and /DE/Rand-to-Best/02. These strategies were tested in five unconstrained continuous benchmark functions with different features and complexity of search space: Rosenbrock, Sphere, Schwefel, Rastrigin, and Griewank. Also, a problem called the economic dispatch problem, whose system comprises 40 generators, was optimized. To compare the strategies, we used a Kruskal-Wallis H-Test. Then we used the pairwise Wilcox test to discover where the differences are. Results have shown that strategies /DE/Best/01 and /DE/Rand-to-Best/01 tend to present the best outcomes. While in the economic dispatch problem, the winner strategy varies as we increase the number of iterations, probably because eventually all strategies can reach a good solution.

In: The Fundamentals of Search Algorithms ISBN: 978-1-53619-007-6
Editor: Robert A. Bohm © 2021 Nova Science Publishers, Inc.

Chapter 1

THE FUNDAMENTALS OF HEURISTIC LOCAL SEARCH ALGORITHMS FOR THE TRAVELING SALESMAN PROBLEM

Weiqi Li[*]
School of Management, University of Michigan-Flint,
Michigan, Flint, US

ABSTRACT

Heuristic local search algorithms are used to find "good" solutions to the NP-hard combinatorial optimization problems that cannot be solved using analytical methods. This chapter discusses the characterization and computation of heuristic local search algorithm for the Traveling Salesman Problem (TSP) from the perspective of dynamical systems. A heuristic local search system is essentially in the domain of dynamical systems. Like many other dynamical systems, a local search system has an attracting property that drives the search trajectories to converge to a small region, called a solution attractor, in the solution space. The study of the solution attractor in the solution space can provide the answer to the question: "Where do the search trajectories go in a heuristic local search system?"

---

[*] Corresponding Author's Email:weli@umich.edu.

The solution attractor collects all locally optimal solutions around the globally optimal solutions. This chapter, using the TSP as the study problem, describes the behavior of search trajectories and properties of the solution attractor in a local search system. Based on these properties of the solution attractor, a novel global optimization algorithm – Attractor-Based Search System (ABSS) – is introduced. This algorithm provides the answer to the question: "How can we use efficient heuristic local search for global optimization?" The ABSS combines a local search process and an exhaustive search procedure. The local search process constructs the solution attractor of the local search system, and the exhaustive search procedure identifies the best solutions in the solution attractor. This chapter describes how the ABSS meets the requirements of global optimization system. The computational complexity of the ABSS for the TSP is also discussed.

Keywords: traveling salesman problem, combinatorial optimization, heuristic local search, global optimization, computational complexity, multimodal optimization, dynamical systems, solution attractor

1. INTRODUCTION

Optimization has been a fundamental tool in all scientific and engineering areas. The goal of optimization is to find the absolutely best set of admissible conditions to achieve our objective in our decision-making process. Researchers have developed many optimization algorithms to solve hard optimization problems. Deterministic approaches such as exhaustive enumeration and branch-and-bound can find exact optimal solutions, but they are very expensive from the computational point of view. The NP-hardness and intractability of many combinatorial optimization problems have led people to employ heuristic local search algorithms and other stochastic optimization algorithms, such as Evolutionary Algorithms (EAs), Particle swarm Optimization (PSO), simulated annealing, and other metaheuristics, to find hopefully good solutions to these problems (Gomey, 2019; Horst and Pardalos, 1995; Korte and Vygen, 2007; Paradimitriou and Steiglitz, 1998; Pardalos et al., 2000; Zhigliavsky and Žillinakas, 2008). The stochastic search algorithms trade in guaranteed correctness of the optimal solution for a shorter computing time. In practice, most stochastic search algorithms have been based on or derived from heuristic local search algorithms (Aart and Lenstra, 2003). Heuristics used in stochastic optimization are functions that help us decide which one of a set of possible solutions is to be selected next (Michalewicz and Fogel, 2002; Rayward-Smith et al., 1996). A local search algorithm iteratively explores the neighborhoods of solutions trying to improve the current solution by a local change. However, the scope of local search is limited by the neighborhood definition. Therefore, local search algorithms are locally convergent. The final solution may deviate from the optimal solution. Such a final solution is called a *locally optimal solution*, denoted as s'. To distinguish from locally optimal solutions, the optimal solution in the solution space is usually called the *globally optimal solution*, denoted as s^*.

The TSP is one of the most intensely investigated problems and often treated as the prototypical combinatorial optimization problem that has provided much motivation for design of new algorithms, development of complexity theory, and analysis of solution space or search space (Applegate et al., 2006; Paradimitriou and Steiglitz, 1998). This chapter uses the TSP as a study problem to describe search behavior of heuristic local search systems. The same concepts and formulation can be used for any combinatorial optimization problem requiring the search of a node permutation in a graph.

Both heuristic local search algorithms and the TSP have been hot research areas for decades, and many aspects of them have been studied. However, there is still a variety of open questions and unsolved issues in these areas. In fact, local search algorithms on the TSP is still a very interesting research problem. The study of local search for the TSP continues to be a vibrant, exciting and fruitful endeavor in computational mathematics, computer science, engineering and artificial intelligence.

One of the fundamental concepts presented in this chapter is that a heuristic local search system has a *solution attractor*, which drives the search trajectories to converge to a small region in the solution space that consists of the most promising solutions to the problem at hand. The goal of studying solution attractor in heuristic local search system is to understand the asymptotic behavior of the search trajectories in a local search system. Important insights can be obtained by observing local search dynamics in the context of attractor, and new search strategies can be designed based on these insights.

This chapter is not prepared for giving a formal theory of the solution attractor. Instead, it describes the concept of the solution attractor informally and presents some experimental results about some important properties of the search trajectories in a local search system without detail theoretical analysis and mathematical proof. In fact, mathematical proof is very difficult in this area. It is hoped that this chapter provide some ideas for future research in both theory and application. The focus of this chapter is trying to answer the question "What is the fundamental nature of the heuristic local search system and how can we use it to perform global optimization search?"

This chapter is organized in the following sections. Section 2 defines the TSP from the global optimization perspective, and then describes the nature of local search system and the concept of solution attractor of a local search system. Section 3 discusses the requirements for a global optimization search system. Section 4 presents a novel global search system for the TSP, and discusses its global search features and computing complexity. The final section concludes the chapter.

2. TRAVELING SALESMAN PROBLEM AND LOCAL SEARCH SYSTEM

2.1. The Traveling Salesman Problem

A problem is the frame into which the solutions fall. By changing the frame, we can change the range of possible solutions. If we want a search algorithm to be a global optimization system for a problem, the problem should be defined from global perspective. The TSP is defined as a complete graph $Q = (V, E, C)$, where $V = \{v_i : i = 1, 2, \ldots, n\}$ is a set of n nodes, $E = \{e(i,j) : i, j = 1, 2, \ldots, n; i \neq j\}$ is an $n \times n$ edge matrix containing the set of edges that connects the n nodes, and $C = \{c(i,j): i,j = 1,2,\ldots,n; i \neq j\}$ is an $n \times n$ cost matrix holding a set of traveling costs associated with the set of edges. The solution space S contains a finite set of all feasible tours that a salesman may traverse. A tour $s \in S$ is a closed route that visits every node exactly once and returns to the starting node at the end. Many real-world optimization problems are inherently multimodal; that is, these problems may contain multiple optimal solutions in their solution spaces. We assume that a TSP instance contains h (≥ 1) optimal tours in the solution space S. We denote $s^* = \min_{s \in S} f(s)$ as a globally optimal tour and S^* as the set of h globally optimal tours. The objective of the TSP is to find all h globally optimal tours in the solution space, that is, $S^* \subset S$:

$$S^* = \arg[\min_{s \in S} f(s)] = [s_1^*, s_2^*, \ldots, s_h^*] \qquad (1)$$

Under this definition, the salesman wants to know what all best alternative tours are available. Finding all optimal solutions is the essential requirement for global search algorithms. In practice, knowledge of multiple optimal solutions is extremely helpful, providing the decision-maker with multiple options. Obviously, this TSP definition is elegantly simple and full of global optimization implications. However, finding all optimal tours is typically a great challenge to the optimization researchers and practitioners.

Usually the edge matrix E is not necessary to be included in the TSP definition because the TSP is a complete graph. However, the edge matrix E is a critical data structure that can help us to understand the search behavior of a local search system and to convert local search into global search in the system. General local search algorithms may not require much problem-specific knowledge in order to generate good solutions. However, it may be unreasonable to expect a search algorithm to be able to solve any problem without taking into account the data structure and properties of the problem at hand. To solve a problem, the first step is to create a manipulatable description of the problem itself. For many problems, the choice of data structure for representing a solution plays a critical role in the analysis of search behavior and design of new search algorithm. For the TSP, a tour can be represented by an ordered list of nodes or an edge configuration of a tour in the edge matrix E, as illustrated in Figure 1. The improvement of the current tour represents the change in the order of the nodes or the edge configuration of a tour.

For the problem setting in all experiments mentioned in the chapter, we generated symmetric TSP instances with n nodes. The cost element $c(i,j)$ in the cost matrix C was assigned a random integer independently drawn from a uniform distribution of the range [1, 1000]. The triangle inequality $c(i,j) + c(j,k) \geq c(i,k)$ was not assumed in the instances. Alghouth this type of problem instances is application-free, it is mathematically interesting and significant. A TSP instance without triangle inequality cannot be approximated within any constant factor. A heuristic local search algorithm usualy performs much worse for this type of TSP instances, which offers a strikingly challenge to solving them (Papadimitriou and Steiglitz, 1977 &1998; Rego et al., 2011; Sahni and Gonzales, 1976; Sourlas, 1986).

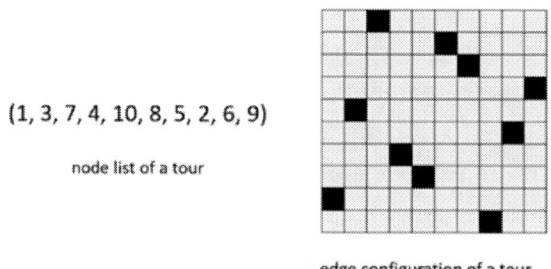

(1, 3, 7, 4, 10, 8, 5, 2, 6, 9)

node list of a tour

edge configuration of a tour

Figure 1. Two representations of a tour: an ordered list of nodes and an edge configuration.

2.2. Heuristic Local Search

The TSP is believed to be an intractable problem. Intractable problems are the problems for which there exists no practically efficient algorithms to solve them (Vavasis, 1995; Papadimitriou and steiglitz, 1977). Only exhaustive search algorithm can provide exact optimal solutions to these problems. However, searching the entire solution space is typically extremely expensive in computing time. Therefore, heuristic search algorithms are designed to find a balance between running time and solution quality (Aarts and Lenstra, 2003). Heuristic algorithms are approximate approaches that do not guarantee that the optimal solutions will be found, but they provide "good enough" suboptimal solutions to the problems at hand in a reasonable amount of time. Many metaheuristic algorithms have been developed or modified to solve the TSP. Nevertheless, the best approximation algorithm for the TSP is still the simple heuristic local search algorithm, which is extremely efficient at finding high-quality solution (Johnson, 1990).

Local search is based on the concept of neighborhood search. A neighborhood of a solution $s_i \in S$, denoted as $N(s_i)$, is a set of solutions that are in some sense close to s_i. For the TSP, a neighborhood of a tour s_i is defined as a set of tours that can be reached from s_i in one single transition. From the view of edge configuration, all tours in $N(s_i)$ are very similar because they share a significant number of edges with s_i. The basic operation of local search is iterative improvement, which starts with an initial tour and searches the neighboroood of the current tour for a better tour. If such a tour is found, it replaces the current tour. The search continues until no improvement can be made. The local search algorithm returns a locally optimal tour.

The bahavior of a local search trajectory can be understood as a process of iterating a search function $g(s)$. We denote s_0 as an initial point of search and $g^t(s)$ as the tth iteration of the search function $g(s)$. A search trajectory $s_0, g(s_0), g^2(s_0), \ldots, g^t(s_0), \ldots$ converges to a locally optimal point s' as its limit, that is,

$$g(s') = g\left(\lim_{t \to \infty} g^t(s_0)\right) = \lim_{t \to \infty} g^{t+1}(s_0) = s' \qquad (2)$$

Therefore, s' is the end point of a search trajectory, and the search trajectory then stays at this point forever.

In a local search system, the globally optimal solution s^* is the target point of search even though we do not know where it is in the solution space.

Due to the constraint of the neighborhood search, a local search trajectory rarely reaches the target point and stops at a locally optimal point. A tour $s_i \in S$ is locally optimal with respect of its neighborhood $N(s_i) \subset S$ if $f(s_i) \leq f(s_j)$ for all $s_j \in N(s_i)$. Different search trajectories will reach differetn locally optimal points because a search trajectory has sensitive dependence on initial point and randomness in the search process. Since all search trajectories have the same target point, they all move towards the same direction and eventually reach the same target region. A single search trajectory converges to a locally optimal point and will stay at that point; and all search trajectories converge to a small region and stay within it. This property of a local search system brings on two assumptions about a local search system: convexity and centrality. The convexity assumption says that the local optimal points are gathered in a relatively small region of the solution space, and the centrality assumption says further that the best of these local optimal points is located centrally with respect to the other local optimal points. If both assumptions are valid, we should expect that all locally optimal points are gathered around the globally optimal point.

In a local search system, there is a great variety of ways to construct initial points, choose candidate moves, and define criteria for accepting candidate moves. Most heuristic local search systems are based on randomization. In this sense, a heurietic local search is a randomized system. There are no two search trajectories that are exactly alike in such a search system. Different search trajectories explore different regions of the solution space and stop at different final points. Therefore, local optimality depends on the initial points, the neighborhood function, randomness in the search process, and time spent on search process. On the other hand, however, a local search system essentially is deterministic and not random in nature. There are some invariant features that are common to all search trajectories. If we visualize the motion of all search trajectories, we will see that the search trajectories go towards the same direction, move closer to each other, and eventually converge into a small region that consists of a small set of distinct points. This behavior reveals a hidden order in a heuristic local search system: locally optimal points appear random only in this small region, not entire solution space. By way of contract, a truly random system will behave in such a way that the search trajectories in solution space wander over the entire space available to the system, and locally optimal points will spread in the entire solution space.

A local search trajectory changes its edge configuration according to the objective function $f(s)$ and its neighborhood structure. Let $N(s_i)$ be the neighborhood of the tour s_i. We assume that all neighborhoods are of equal size, i.e., $N(s_i) = \Theta$, for all $s_i \in S$. If s_i is the current tour at time t, and the search trajectory chooses a neighbor $s_j \in N(s_i)$ at random, the next tour s_{t+1} is determined by the following rule:

$$\begin{cases} s_{t+1} = s_j & \text{if } f(s_j) < f(s_i) \\ s_{t+1} = s_i & \text{otherwise} \end{cases} \qquad (3)$$

Therefore, the probability of the next tour depends partially on the outcome of the previous tour. According to the theory of Markov chain (Gagniuc, 2017), a search trajectory generates a discrete time series of Markov chain. Each next tour in the time series is statistically dependent on the edge configurations in the tour sequence. In certain way, the search trajectories can be studied using the theory of finite Markov chains. In this sense, the evolution of edge configuration of a search trajectory can be viewed as the time evolution of a Markov random field. When the matrix E collects the edge configurations of the tours from K search trajectories, it becomes a collection of Markov chains and itself is also a Markov chain. The final edge configuration of the matrix E is the result of the Markov chain. One essential property of Markov chains is stationarity: under certain conditions, on the transition probabilities associated with a Markov chain, there exists a unique stationary distribution (Meyn and Tweedie, 2009). If we use this Markov chain property to study the convergence of edges in the matrix E, we will see that the hit-frequency values in the edges of the matrix E provide a global description of the asymptotic behavior of the local search system. The final distribution of the hit-frequency on the edges is statistically fixed. Therefore, the set of K search trajectories generates a unique final edge configuration in the matrix E. In the frame of the matrix E, a local search system is a system with 4-tutple (W, M_W, M_A, g), where W

is the set of all edges, M_W is the edge configuration of K initial tours that covers all edges, M_A is the edge configuration of the K locally optimal tours, and g is the search function or transition function. In a local search system, g is not a well-defined function but embodies the principle of determinism. Each transition is a nondeterministic operation and may correspond more than one possible next tours from the current tour. However, the result of the transitions for K search trajectories are always the same. The final edge configuration of the matrix E is not sensitive to the selection of K search trajectories. That is, for all different M_W, we always obtain the same M_A.

In our experiments, we used the 2-opt method, a general local search technique applicable to various NP-hard problems. The 2-opt neighborhood can be characterized as the neighborhood that induces the greatest correlation between function values of neighboring tours, because neighboring tours differ in the minimum possible four edges. Along the same reasoning lines, the 2-opt may have the smallest expected number of locally optimal points (Savage, 1976). The local search process randomly selected a solution in the neighborhood of the current solution. A move that gave the first improvement was chosen. The great advantage of the first-improvement pivoting rule is to produce randomized locally optimal points. The software programs for the experiments were written using different programming languages and were run in PCs with different versions of Window operating systems.

2.3. Solution Attractor of Heuristic Local Search System

Heuristic local search algorithms are essentially in the domain of dynamical systems (Li, 2005). A heuristic local search algorithm is a discrete dynamical system, which has a solution space S (the state space), a set of times T (search iterations), and a search function $g: S \times T \to S$ that gives the consequents to a solution $s \in S$ in the form of $s_{t+1} = g(s_t)$. A *search trajectory* is the sequence of states of a local search system at successive time-steps, which represents the part of the solution space searched by this search trajectory (Li and Feng, 2013). Questions about the behavior of the local search system over time are actually the questions about its search trajectories. The most basic question about the local search trajectories is "Where do they go in the solution space and what do they do when they get there?"

The attractor theory of dynamical systems is a natural paradigm that can be used to describe the search behavior of a heuristic local search system. The theory of dynamical systems is an extremely broad area of study. A dynamical system is a model of describing the temporal evolution of a system in its state space. The goal of dynamical system analysis is to capture the distinctive properties of certain points or regions in the state space of a given dynamical system (Allifood, et al., 1997; Auslander et al., 1964; Dénes and Makey, 2011; Fogedby, 1992; Milnor, 2010; Ruelle, 1981). The theory of dynamical systems has discovered that many dynamical systems exhibit attracting behavior in the state space (Brin and Stuck, 2016; Brown, 2018). In such a system, all initial states tend to evolve towards a single final point or a set of points. The term *attractor* is used to describe this single point or the set of points in the state space. An attractor is invariant under the dynamics, towards which the states in a given basin of attraction asymptotically approach in the course of dynamic evolution (Alligood et al., 1997; Auslander et al., 1964; Milnor, 1985&2010). The basin of attraction is defined as the set of initial points whose trajectories tend towards the attractor. The attractor theory of dynamical systems describes the asymptotic behavior of typical trajectories in the dynamical system (Buescu, 1991; Collet and Eckmann, 1980; Dénes and Makey, 2011; Milnor, 2010; Ruelle, 1981). Therefore, the attractor theory of dynamical systems provides the theoretical foundation to study the search behavior of a local search system.

In a local search system for the TSP, no matter where we start K search trajectories in the solution space, all search trajectories will converge to a small region in the solution space for a unimodal TSP instance or h small regions for a h-modal TSP. We call this small region a *solution attractor*, denoted as A, of the local search system for a given TSP instance. Therefore, the solution attractor of a local search system for the TSP can be defined as an invariant set $A \subset S$ consisting of all locally optimal tours and the globally optimal tours. A single search trajectory typically converges to either one of the points in the solution attractor. A search trajectory that is in the solution attractor will remain within the solution attractor forward in time. Because a globally optimal tour s^* is a special case of locally optimal tours, the globally optimal tour is undoubtedly embodied in the solutin attractor, that is, $s^* \in A$. For a h-modal TSP instance, a local search system will generate h solution attractors $(A_1, A_2, \ldots, A_h)$ that attract all search trajectories. Each of the solution attractors has its own set of locally optimal tours, surrounding a globally optimal point s_i^* ($i = 1,2, \ldots, h$). A particular search trajectory will converge into one of the h solution attractors. The set of locally optimal tours generated by K search trajectories will be distributed to these solution attractors. According to dynamical systems theory (Milnor, 2010), the closure of an arbitrary union of attractors is still an attractor. Therefore, the solution attractor A of a local search system for a h-modal TSP is a complete collection of h solution attractors $A = A_1 \cup A_2 \cup \ldots \cup A_h$.

The concept of solution attractor of local search system is based on asymptotic behavior of search trajectories from every choice of initial point in the solution space, which describes where the search trajectories actually go and where their final points actually stay in the solution space. Figure 2 visually summaries the concepts of search trajectories and solution attractors in a local system for a multimodal optimization problem. Illustrating search trajectories, solution attractors, and the solution space as a three-dimensional object is a valid metaphor for understanding how a local search system might proceed, how search trajectories converge, and how solution attractors are formed.

In summary, let $g(s)$ be a search function in a local search system for the TSP, we say that a compact set $A \subset S$ is a solution attractor for the search function $g(s)$ if it satisfies the following properties (Li, 2005; Li and Feng, 2013; Li and Li, 2019):

- Convexity, i.e., $\forall s_i \in S, g^t(s_i) \in A$ for sufficient long t;
- Centrality, i.e., the globally optimal tour s^* is located centrally with respect to the other locally optimal tours in A;
- Invariance, i.e., $\forall s' \in A, g^t(s') = s'$ and $g^t(A) = A$ for all time t;

- Irreducibility, i.e., the solutiona ttractor A contains a limit number of invariant locally optimal tours.

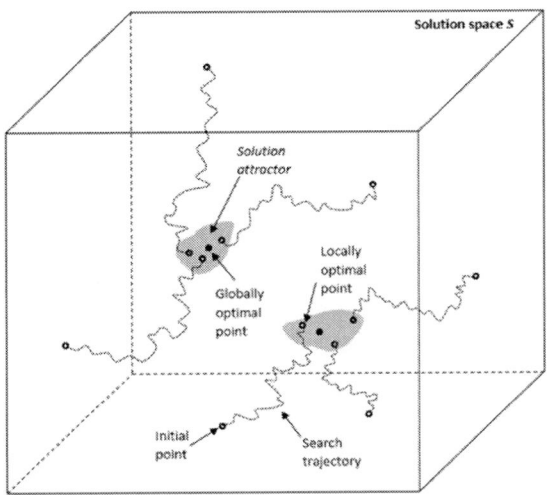

Figure 2. Illustration of the concepts of search trajectories and solution attractors in a local search system for a multimodal optimization problem.

2.4. The Characteristics of the Edge Matrix E

Observing the behavior of search trajectories and envisioning the solution attractors in a local search system can be quite challenging. The edge matrix $E = \{e(i,j)\}_{n \times n}$ in the TSP definiton is a natural data struccture that can help us to trace the movement of the search trajectories and understand the dynamics of a local search system. In the TSP, an edge $e(i,j)$ is the most basic element of a tour, but contains the richest information about $(n-2)!$ tours that go through it. Essentially, the nature of local search for the TSP is an edge-selection process: the preservation of good edges and the rejection of bad edges according to the objective function $f(s)$. Each edge has an implicit probability to be selected by a locally optimal tour. A better edge has higher probability to be included in a locally optimal tour. Therefore, the edges in the matrix E can be divided into three groups: G-edges, globally superior edges and bad edges. A G-edge is the edge that is included in a globally optimal tour. A globally superior edge is the edge that occurs in many or all locally optimal tours. Although each of these locally optimal tours selects this edge based on its own neighborhood function and search trajectory, the edge is globally superior since the edge is selected by these individual tours from different search trajectories going through different search regions. The globally superior edges have higher probability to be selected by a locally optimal tour. All G-edges are globally superior edges and can be treated as a subset of the globally superior edges. The edges that are discarded by all search trajectories or selected by only few locally optimal tours are bad edges. A bad edge is impossible to be included in a globally optimal tour. When a search trajectory stops at a locally optimal tour, this tour consists of some G-edges, some globally superior edges and a few bad edges.

The edge matrix E can follow the "footprints" of search trajectories in the search system to display the dynamics of the search system. The changes of the edge configuration in the matrix E represent the transformations of the search trajectories in a local search system. When all search trajectories reach their end points, the final edge configuration of the matrix E represents the final state of the local search system. The matrix E acts as a coding system that can encapsulate all information about the convergent behavior of search trajectories in a local search system for the TSP. For a tour s_i, we define an element $e(i,j)$ of the matrix E,

$$e(i,j) = \begin{cases} 1 & \text{if the element } e(i,j) \text{ is in the tour } s_i \\ 0 & \text{otherwise} \end{cases} \quad (4)$$

Then the hit-frequency value e_{ij} in the element $e(i,j)$ is defined as the number of occurrence of the element in K tours, that is,

$$e_{ij} = \sum_{k=1}^{K} e(i,j)^k \quad (5)$$

When K search trajectories reach their end points, the value $(e_{ij} + e_{ji})/K$ can prepresent the probability of the edge $e(i,j)$ being hit by a locally optimal tour. Figure 3 shows a simple example of hit-frequency values in

The Fundamentals of Heuristic Local Search Algorithms ...   15

the matrix E, in which five tours $s_1 = (1,2,3,4,5), s_2 = (2,4,5,3,1), s_3 = (4,5,3,2,1), s_4 = (5,1,4,2,3)$ and $s_5 = (3,2,4,5,1)$ are recorded.

	1	2	3	4	5
1	0	2	1	2	0
2	1	0	2	2	0
3	1	2	0	1	1
4	0	1	0	0	4
5	3	0	2	0	0

Figure 3. An example of hit-frequency values in the matrix E.

We can use graphical technique to observe the convergent behavior of search trajectories through the matrix E. The hit-frequency value e_{ij} can be easily converted into a unit of half-tone information in a computer, a value that we interpret as a number H_{ij} somewhere between 0 and 1. The value 1 corresponds to black color, 0 to white color, and any value in between to a gray level. Let K be the number of tours, the half-tone information H_{ij} on a computer screen can be represented by the hit-frequency e_{ij} in the element $e(i,j)$ of the matrix E:

$$H_{ij} = \frac{e_{ij}}{K} \qquad (6)$$

Figure 4 illustrates a simple example of visualization that shows the convergent behavior of local search system for a 50-node instance. The local search system performs 100 local search trajectories concurrently. Figure 4(a) shows the image of the edge configuration of 100 random initial tours. Since each element of the matrix E has equal chance to be hit by these initial tours, almost all elements are hit by these initial tours, and all elements have very low half-tone values, ranging from 0.00 to 0.02. When the local search system starts searching, the search trajectories constantly change their edge configurations, and therefore the colors in the elements of the matrix E are changed accordingly. Some elements become white color and other elements have darker gray color. As the search continues, more elements become white (i.e., they are discarded by all search trajectories), and other elements become darker (i.e., they are selected by more search trajectories). When all search trajectories reach their end points (locally optimal points), the colored elements represents the edge configuration of the final state of the search system, that is, the edge configuration of the solution attractor. Figure 4(b) and (c) show the images of edge configuration of the matrix E when all search trajectories completed 2000 iterations and 5000 iterations, respectively. At 5000th iteration, the range of half-tone values in the elements of the matrix E is from 0.00 to 0.42. The value 0.42 means that 42% of search trajectories selected this element. Many elements become white color (having zero half-tone value).

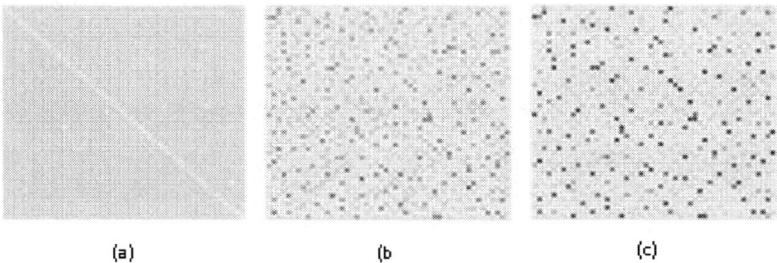

Figure 4. Visualization of the convergent dynamics of edge configuration of the matrix E. (a) the half-tone image of the edge configuration of 100 initial tours, (b) and (c) the half-tone images of edge configurations when the search trajectories are at 2000th and 5000th iteration, respectively.

This simple example has great explanatory power about the global dynamics of the local search system for the TSP. As local search continues, the number of edges hit by the search trajectories becomes smaller and smaller, and better edges are hit by more and more search trajectories. This edge-convergence phenomenon means that all search trajectories are moving closer and closer to each other, and they are increasingly similar in their edge configurations. The distribution of the hit-frequency values in the elements of the matrix E describes the globally asymptotic behavior of the local search system.

When all search trajectories reach their end points – the locally optimal tours, the edge configuration of the matrix E will become fixed. This fixed edge configuration contains two groups of edges: the edges that are not hit by any of K locally optimal tours (non-hit edges) and the edges that are hit by at least one of the locally optimal tours (hit edges). The set of the hit edges forms the edge configuration of the solution attractor, which includes all globally superior edges (including all G-edges) and some bad edges. Figure 5 shows the edge grouping in the matrix E when all search trajectories stop at their end points.

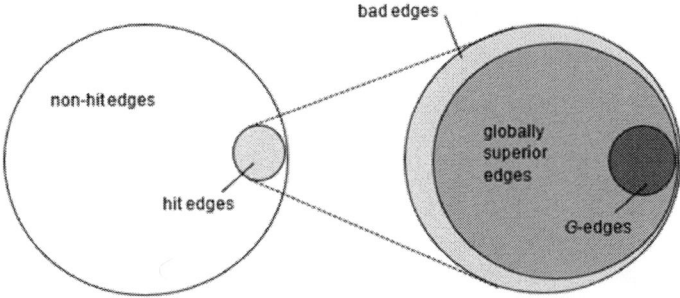

Figure 5. the grouping of the edges in the matrix E.

It is easily verified that under certain conditions, a local search system asymptotically converges to the set of the globally optimal solutions S^*, i.e.,

$$\lim_{K \to \infty} P[S^* \subset S] = 1 \qquad (7)$$

which means that the probability of finding all globally optimal tours is one if the number of search trajectories is unlimited. However, the required search effort may be very huge – equivalent to enumerating all tours in the solution space. Now the question is "How can we construct the edge configuration of the solution attractor without large number of search trajectories?" One of the fundamental theories that can help us to find the answer to this question is the information theory (Shammon, 1948). According to the information theory, each solution point contains some information about its neighboring solutions that can be modelled as a function, called *information* or *influence function*. The influence function of the ith solution point in the solution space S is defined as a function $\Omega_i: S \to \Re$, such that Ω_i is a decreasing function of the distance from a solution point to the ith solution point. The notion of influence function has been used extensively in datamining, data clustering, and pattern recognition. Therefore, the edge matrix E has another important feature: the union of the edge configurations of a set of tours contains information about many other tours because one tour shares its edges with many other tours. Figure 6 uses a simple example to illustrate this feature. Suppose we have three tours s_1, s_2 and s_3 listed in Figure 6(a). We record their edge configurations in the matrix E, and the elements that are hit by these tours are marked with *, as shown in Figure 6(b). The union of the edge configurations of the three tours may contain information about the edge configurations of other tours, as illustrated in Figure 6(c). From the edge configuration shown in Figure 6(b), we can easily identify another tour s_4, as shown in Figure 6(d).

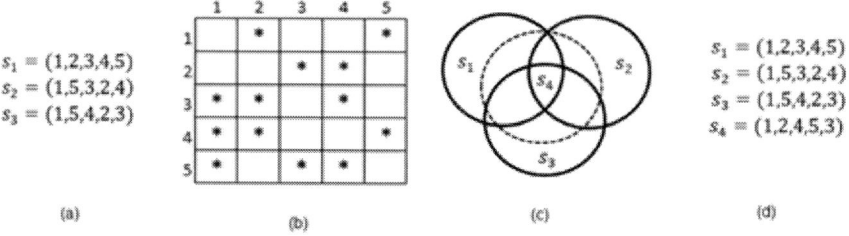

Figure 6. (a) three tours, (b) the edge configurations of these tours stored in the matrix E, (c) the union of the edge configurations contains information about other tours, (d) a new tour s_4 found in the edge configuration of the matrix E.

The matrix E consists of $n(n-1)$ elements (excluding the diagonal elements). When we randomly generate a tour and record its edge configuration in the matrix E, n elements of the matrix E will be hit by this tour. If we generate more random tours and record their edge configurations in the matrix E, more elements will be hit. We define K as the number of random initial tours, whose edge configurations together will hit all elements of the matrix E. We know that all elements of the matrix E represent all combinatorial possibilities in the solution space S. Therefore, K is the number of search trajectories such that the union of edge configurations of their initial tours covers the entire solution space S. In our experiments, we found that the edge configurations of at most $6n$ randomly-generated tours can guarantee to hit all elements of the matrix E. From the tour perspective, $K = 6n$ random tours are only a small set of of the tours in the solution space S; but from the edge-configuration perspective, the union of the edge configurations of $6n$ random tours represents the edge configurations of all tours in the solution space S. It reveals an amazing fact: the union of edge configurations of only $6n$ randomly-generated tours contains the edge configurations of all $(n-1)!/2$ tours in the solution space. It reflects the combinatorial nature of the TSP: the tours in the solution space are formed by different combinations of the edges. Decades ago, Lin (1965) used this combinatorial property to design a new search strategy for the TSP. He observed the correlation between local optimal tours and proposed a strategy that obtains several local optimal tours and then identify edges that are common to all of them. These edges are then fixed and used to recombine to find more other locally optimal tours.

Based on the information theory, we can assume that the union of the edge configurations of the locally optimal tours generated by K search trajectories typically contain the information about the edge configurations of all other locally optimal tours and the globally optimal tours. The K search trajectories start from the entire solution space and explore different regions to collect all globally superior edges, therefore, the union of edge configurations of their final tours is not just a simple union of the edge configurations of the K locally optimal tours, it also includes the edge configurations of all other locally optimal tours.

For a set of K search trajectories to be converging, they must be getting closer and closer to each other. That is, these search trajectories become increasingly similar in their edge configurations by sharing more and more edges with each other. As a result, the edge configurations of the K search trajectories converge to a small set of edges that contains all globally superior edges. Let W denote total number of edges in the matrix E, $\alpha(t)$ the number of the edges that are hit by all K search trajectories at time t, $\beta(t)$ the number of the edges that are hit by one or some of K search trajectories, and $\gamma(t)$ the number of edges that have no hit at all, then at any time t, we have

$$W = \alpha(t) + \beta(t) + \gamma(t) \tag{8}$$

For a given TSP instance, W is a constant value $n(n-1)/2$ for a symmetric instance or $n(n-1)$ for an asymmetric instance. During the local search process, the values for $\alpha(t), \beta(t)$ and $\gamma(t)$ are constantly changing. One can expect that, as local search process consinues, the values for both $\alpha(t)$ and $\gamma(t)$ will increase, and the value for $\beta(t)$ will decrease. Our experiments confirmed this inference about $\alpha(t), \beta(t)$ and $\gamma(t)$. Figure 7 illustrates the patterns of $\alpha(t), \beta(t)$ and $\gamma(t)$ curves generated from our experiments. These curves cannot increase or decrease forever. At certain point of time, they will become stable, and the values for $\alpha(t), \beta(t)$ and $\gamma(t)$ will reach constant values, that is,

$$W = \lim_{t \to \infty} \alpha(t) + \lim_{t \to \infty} \beta(t) + \lim_{t \to \infty} \gamma(t) = A + B + \Gamma \tag{9}$$

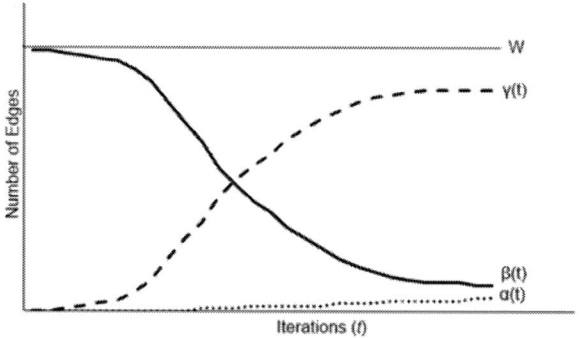

Figure 7. The $\alpha(t), \beta(t)$ and $\gamma(t)$ curves with search iterations.

Therefore, the distribution of the hit-frequencies in the edges of the matrix E eventually becomes fixed. At that time, the edge configuration of the matrix E is a real edge configuration of the solution attractor. The matrix E captures the evolution of the local search trajectories over time and reveals the global dynamics of the local search system: *the edge configurations of all search trajectories will converge to a small fixed set of edges*. Our experiments also found that, for unimodal TSP instances, the ratio $\gamma(t)/W$ could approach to 0.70 quickly for different sizes of TSP instances. For multimodal TSP instances, this ratio depends on the number of the globally optimal points. However, the size of the union of the multiple solution attractors is still very small.

Different sets of K search trajectories will generate different final edge configurations in the matrix E. However, the difference is very small and exists only in the bad-edge set. The underlying edge configuration of the solution attractor is structurally stable because small changes in the set of bad edges leave the edge configuration of the solution attractor qualitatively unchanged. Suppose that, we start the local search from a set of K initial points and obtain the edge configuration M_a in the matrix E when the local search process is terminated. Then we start the local search process again from a different set of K initial points and obtain a little different edge configuration M_b. Which edge configuration truly describes the edge configuration of the solution attractor? Actually, they are structurally equivalent because they are different only in the set of bad edges. In other words, M_a and M_b are topologically conjugate, each precisely replicated the dynamical properties of the other, thus the two edge configurations can be considered equivalent. Therefore, a local search system actually is a deterministic system: although a single search trajectory appears stochastic, all search trajectories from different initial points will be always trapped into the same small region by a hidden order in the local search system and the final edge configuration will always converge to the same set of the globally superior edges.

In our study, we use the change in the edge configuration of the matrix E to measure the convergence of the search trajectories. Therefore, the convergence in our study is defined as the state of aggregated edge configuration of K search trajectories in the matrix E. In the local search process, K search trajectories collect all available topology information about the quality of the edges from their search experiences and record the information in the matrix E. The change in the edge configuration of the matrix E fully reflects the real search evolution of the local search system.

A state of convergence is achieved once no any more local search trajectory can change the edge configuration of the matrix E. That is, the edge configuration of the matrix E becomes fixed, which is the final global structure of a search space. Convergence of the search trajectories and stability of the solution attractor are two desirable properties of a local search system. That is, all local search trajectories will converge to the solution attractor and remain there.

In summary, we assume a TSP instance Q has a solution space S that contains $h (\geq 1)$ globally optimal tours $S^* = [s_1^*, s_2^*, \ldots, s_h^*]$, and correspondingly there exist h sets of G-edges $G = G_1 \cup G_2 \cup \ldots \cup G_h$. A local search system for Q will generate h solution attractor $A = A_1 \cup A_2 \cup \ldots \cup A_h$ that attract all search trajectories. Each of the solution attractors has its own set of locally optimal tours, surrounding a globally optimal point $s_i^* (i = 1, 2, \ldots, h)$. A particular search trajectory will converge into one of these solution attractors. The set of locally optimal tours generated by all search trajectories will be distributed to these solution attractors. Each of the solution attractors has its own edge configuration. Two solution attractors may intersect each other in edges. The edge configuration of the solution attractor A is the union of the edge configurations of the h solution attractors. The final edge configuration of the matrix E represents the edge configuration of A with two properties;

- It contains a complete collection of solution attractors, i.e., $A = A_1 \cup A_2 \cup \ldots \cup A_h$;
- It contains a complete collection of G-edges; i.e., $G = G_1 \cup G_2 \cup \ldots \cup G_h$.

The ability of K search trajectories to explore the edge space in the matrix E and thus collect all G-edges indicates that the solution attractor A contains all globally optimal tours. The global convergence and deterministic property of the search trajectories make the local search system always converge to the same solution attractors and the edge configurations of the search trajectories converge to the same set of globally superior edges.

3. THE REQUIREMENTS FOR A GLOBAL SEARCH SYSTEM

Global optimization is concerned with the computation and characterization of globally optimal solutions in the solution space (Horst and Pardalos, 1995). The task of a global optimization system is to find all absolutely best solutions in the solution space. Therefore, global optimization can be defined as the problem of finding the whole set of optimal solutions $S^* \subset S$, and any optimal solution $s^* \in S^*$ optimizes the objective function $f(s)$ and satisfies a set of restrictions. There are two major tasks performed by a global optimization system : (1) finding all globally optimal points in the solution space, and (2) making sure that they are globally optimal.

Usually when we have a new large problem instance, we do not know how many optimal solutions exist in the solution space. For NP-hard combinatorial problems, there is no well-developed theory or analysis tool available to answer this question. The number of the optimal solutions for a given combinatorial optimization problem can be known only after we check all possible solutions in the solution space. The TSP can be solved exactly using exhaustive search and branch-and-bound search techniques. However, for large TSP instance, no exact algorithm is feasible. Optimization practitioners have to utilize heuristic search algorithms such as random-restart hill-climbing, simulated annealing and many other meta-heuristic algorithms (Michalewicz and Fogel, 2002; Reeves, 1993). Most such search algorithms are designed to locate any one of the optimal solutions in the solution space. In order to find all optimal solutions, we need a search algorithm that converges not only in value but also in solution. *Convergence in value* means that a search system can eventually find any one of the optimal solutions. *Convergence in solution* means that the search system can identify repeatedly all optimal solutions. Always finding the same set of optimal solutions actually is the fundamental requirement for global optimization systems.

So far, we do not have any effective and efficient global search algorithm to solve hard combinatorial optimization problems. Currently, evolutionary algorithms (EAs) are widely-used techniques in global optimization because their population-based and parallel search capability make them look particularly effective in finding multiple optimal solutions. However, the original forms of EAs are developed for locating single optimal solution, because the natural tendency of EAs will always converge to one best solution or sub-optimal solution (Masel, 2011). To overcome this problem, a variety of dedicated techniques, commonly known as niching methods, have been developed and incorporated in EAs to locate and maintain multiple good solutions (Mahfoud, 1995). The fundamental assumption behind the niching techniques is that a niching technique is able to determine the class memberships for all points in the solution space, and an element in a given class is always closer to every element of its own class than any element of another class.

However, the modern global search algorithms do not provide an answer to the fundamental question of how to recognize the optimal solutions. This critical question is still open. A significant difficulty with global optimization is the lack of practical criteria that decides when a locally optimal solution is a globally optimal one. What is the necessary and sufficient conditions for a feasible point to be globally optimal point? From a mathematical point of view, the general global optimization problem is essentially unsolvable, due to a lack of mathematical conditions characterizing the global optimum. The globally optimal points s^* can only be obtained by a deterministic algorithm. So far, the only deterministic method that can determine the globally optimal points is the exhaustive search. However, this method is typically extremely expensive. Therefore, people introduce stochastic elements into the deterministic algorithm. In such a way, the deterministic guarantee that the global optimum can be found is relaxed into a confidence measure. Stochastic global optimization algorithms contain a random element in the choice of the next computing step. Stochastic methods can be used to assess the probability of having obtained the global optimum. In order to recognize the optimal solutions, any optimality criterion must be based on information on the global behavior of the search system. An optimality criterion is essentially an equivalent restatement of the definition of optimality. Therefore, a global optimization system should meet the following requirements :

- The search behavior of the system should be globally convergent.
- The search system should be deterministic and have a rigorous guarantee for finding all globally optimal solutions without excessive computational burden.
- The search system should use an optimality criterion to make sure the best solution found is the globally optimal solution.

4. THE ATTRACTOR-BASED SEARCH SYSTEM (ABSS)

4.1. The ABSS for the TSP

Development of efficient global optimization systems is much more difficult than that of the local optimizaton methods. For global optimization, the central idea should be globally descent. In addition, the rationality of global search strategy for finding global optimum should be justified. The converging behavior of the search trajectories and the stability of the solution attractor in local search systems are of great interest. The definition of solution attractor not only defines what a solution attractor *is*, but also describes what it *does* and what we can *do with it*. The solution attractor tells us in which region a globally optimal point is located. The concept of the solutioin attractor of local search system not only helps us understand the behavior of local search trajectories, but also offers an important method to solve the TSP efficiently with global optimality guarantee.

Figure 8 presents the ABSS for the TSP. In this algorithm, Q is a TSP instance with the edge matrix E and cost matrix C. At beginning of search, E is initialized by assigning zeros to all elements. The function InitialTour() constructs an initial tour s_i using any tour-construction technique. The function LcalSearch() takes s_i as input, performs local search using any type of local search technique, and returns a locally optimal tour s_j. The function UpdateE() updates E by recording the edge configuration of tour s_j into E. K is the number of search trajectories. After the edge configurations of K locally optimal tours are stored in E, the

ExhaustedSearch() function completely searches E using the depth-first tree search technique, a simple recursive search method that traveses a directed graph starting from a node and then searches adjacent nodes recursively. Finally, the ABSS outputs a set of all best tours S^* found in the edge configuration of E. The search strategy in the ABSS is straghtforward: generating K locally optimal tours, storing their edge configurations in the matrix E to form the edge configuration of the soluion attractor, and then identifying the best tours by evaluating all tours represented by the edge configuration of the solution attractor. The ABSS is a simple efficient computer program that can solve the TSP effectively. This search system shows strong features of effectiveness, flexibility, adaptability and scalability. The computationl model of the ABSS is inherently parallel, facilitating implementations on concurrent processors. It can be implemented in many different ways: series, parallel, distributed, or hybrid.

```
1     ABSS_Algorithm(Q)
2     begin
3        Initialize E;
4        NumberOfTrajectories = 0;
5        repeat
6           s_i = InitialTour();
7           s_j = LocalSearch(s_i);
8           E = UpdateE(E, s_j);
9           NumberOfTrajectories = NumberOfTrajectories + 1;
10       until NumberOfTrajectories = K
11       S* = ExhaustedSearch(E);
12    end
```

Figure 8. The ABSS algorithm for the TSP.

The philosophy behind the ABSS is that: if we do not know where the globally optimal point is, we can try to find the small region that is known to contain this optimal point and then search that region thoroughly. Imagine that searching for the optimal solution in the solution space is like treasure hunting. We are trying to hunt for a hidden treasure in the whole world. If we are "blindfolded" without any guidance, it is silly idea to search every single square inch of the extremely large space. We may have to perform a random search process, which is usually not effective. However, if we are able to use various clues to locate a small village where the treasure was placed, we will then directly go to that village and search every corner of the village to find the hidden treasure. The ABSS uses the same strategy.

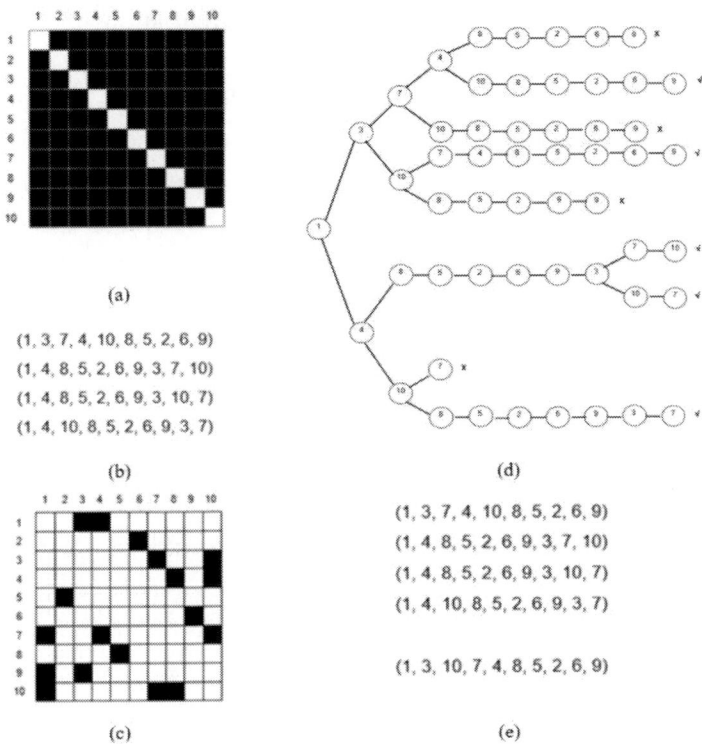

Figure 9. A simple example of the ABSS search. (a) the edge configuration of 60 initial tours, (b) four distinct locally optimal tours, (c) the edge configurations of the four locally optimal tours, (d) the depth-first tree search on the edge configuration of the solution attractor, and (e) all five tours found in the solution attractor.

Figure 9 uses a 10-node instance as an example to illustrate how the ABSS works. We randomly generate $K = 6n = 60$ initial tours, which edge configurations black all elements of the matrix E, as shown in Figure 9(a). It means that these 60 initial tours hit all 45 edges that represent all 181440 tours in the solution space. We let each of the search trajectories run 5000 iterations and obtain 60 locally optimal tours. However, due to the small size of the instance, most locally optimal tours have identical edge configurations. Among the 60 locally optimal tours, we find only four distinct locally optimal tours as shown in Figure 9(b). Figure 9(c) shows the union of the edge configurations of the 60 locally optimal tours, in which 18 edges are hit. Then we use the depth-first tree search, as illustrated in Figure 9(d), to identify all five tours in the edge configuration of the solution attractor, which are listed in Figure 9(e). In fact, one of the five tours is the globally optimal tour. This simple example indicates that (1) local search trajectories converge to small set of edges, and (2) the union of edge configurations of K locally optimal tours is not a countable union of the edge configuration of the these tours, it also include the edge configurations of other locally optimal tours. It is one of the critical features of the ABSS.

The ABSS consists of two search phases: local search phase and exhaustive search phase. The task of the local search is to identify the village where the hidden treasure was placed (i.e., the solution attractor), and the task of the exhaustive search is to find the treasure (i.e., the globally optimal points). In the local search phase, the local search trajectories actually perform the task of pruning out the edges that cannot possibly be G-edges. When one edge is discarded by all search trajectories, all tours that pass the edge will be removed from the search space for the exhaustive search. When majority of the edges are removed, most connections between nodes are also removed. When the edge configurations of K locally optimal tours are recorded into the matrix E, the hit-frequency values in the edges of the matrix E is the information about suitability of the edges for global optimality. This action significantly reduces the number of combinatorial branching possibilities at each node. Therefore, huge number of possible tours in the solution space are excluded in the search space for the exhaustive search phase. The local search trajectories cut off huge number of useless branches from the search tree. The K search trajectories start from a complete set of edges that represents all combinatorial branching possibilities in the solution space and converge to a small set of edges that represent a limited number of tours in the solution attractor. However, it is still difficult to identify G-edges among the globally superior edges directly, we should perform an exhaustive search in the solution attractor to find them. Since local search phase has significantly reduced the computational complexity of the search space, the exhaustive search phase becomes feasible.

4.2. Global Optimization Features of the ABSS

How does the ABSS meet the global optimization requirements? In section 2, we have discussed that the edge configurations of K search trajectories converge globally to a small fixed set of edges. The ABSS combines beautifully two crucial aspects in search: exploration and exploitation (or diversified search and intensified search). In the local search phase, K search trajectories explore the full solution space to identify the edge configuration of the solution attractor (i.e., the location of the solution attractor in the solution space). These K search trajectories are independently and individually executed, and therefore they create and maintain diversity from begining to the end. The local search phase is a randomized process due to randomization in the local search function $g(s)$. Therefore, the K search trajectories actually perform the Monte Carlo simulation to sample locally optimal tours. A Monte Carlo simulation is defined as a simulation used to model the probability of different outcomes in a process that cannot easily be predicted due to intervention of random variables (Hastings, 1970). The essential idea of Monte Carlo method is using randomness to solve problems that might be deterministic in principle (Hastings, 1970, Kolokoltsov, 2010; Kroese et al., 2011). In the ABSS, K search trajectories start a sample of initial points from a uniform distribution over the solution space S, and through the randomized local search process, generate a sample of locally optimal points uniformly distributed in the solution attractor A. Therefore, the edge configuration of the solution attractor is actually constructed through this Monte Carlo sampling process.

The objective function $f(s)$ is the key force that directs the search trajectories to the solution attractors. The matrix E is a critical data structure that transforms local search behavior of the individual search trajectories into a collective global search behavior of the system. Each time when a local search trajectory finds a better tour and updates the edge configuration of the matrix E, the conditional distribution on the edges are updated and more values are attached to the globally superior edges. Let W be the complete set of the edges in the matrix E, g the local search funciton, and W_A the set of edge in the edge configuration of the solution attractor such that $g(W)$ is contained in the interior of W. Then the intersection W_A of the nested sequence of sets is

$$W \supset g(W) \supset g^2(W) \supset \cdots \supset g^t(W) \supset \cdots \supset W_A \qquad (10)$$

and the intersection W_A is always invariant, that is, $g^t(W_A) = W_A$. As a result, the edge configurations of K search trajectories converge to a small fixed set of edges that contains the edge configurations of all locally optimal tours.

A great challenge for local search systems is to set up a practical optimality criterion used to decide whether a locally optimal tour is also a globally optimal tour. For a TSP instance, there exists unknown number of optimal tours in its solution space. In mathematics, it is still difficult to find a general way to characterize the global optimality for combinatorial optimization problems. Selecting the best tour among a set of tours and knowing it is the best is still full challenge of the TSP. Of course, an exhaustive algorithm that sorts through all tours in the solution space meets the challenge. But it lacks practical efficiency. We know that for a given s^* to be a minimum, a necessary condition is $\lim_{t \to \infty} g^t(s^*) = s^*$, i.e., s^* is a stationary point and further search iteration will not change its edge configuration. What is the additional condition that insures the stationary point s^* is a global minimum? Obviously, the additional condition is that the search system should provide globality of search. Globality of search means that the search trajectories are spread all over the entire solution space S. Therefore, the necessary and sufficient condition for s^* being a global minimum in S is: $\forall s \in S, f(s^*) \leq f(s)$. Any optimality criterion must be based on global information of the search system.

In the ABSS, what are the necessary and sufficient conditions for a locally optimal tour to be a globally optimal tour in the solution space? The

ABSS uses a simple and practical optimality criterion: the best tour among the set of all locally optimal tours is the globally optimal tour. This criterion describes how the ABSS models and solves the TSP. The ABSS find the globally optimal tours through the two-phase search algorithm: the local search phase finds all locally optimal tours over the entire solution space and then the exhaustive search phase examines all locally optimal tours to identify the best ones. When K search trajectories reach their final points, the union of their search spaces should cover the entire solution space. In other words, any tour in A is a locally optimal s_i', such that $f(s_i') \leq f(s)$ for all $s \in (S - A)$. Each search trajectory passes through many neighborhoods on its way to the end point. For any tour s_i in the TSP, the size of $N(s_i)$ is greater than $\left(\frac{n}{2}\right)!$ (Savage, 1976). Let $N(s_i')$ denote the neighborhood of the final point s_i' of the ith search trajectory and $\Omega N(s_{tran})_i$ as the union of the neighborhoods of all transition points of the search trajectory, then we can believe that the search space covered by K search trajectory is

$$N(s_1') \cup \Omega N(s_{tran})_1 \cup N(s_2') \cup \Omega N(s_{tran})_2 \cup \ldots \cup N(s_K') \cup \Omega N(s_{tran})_K = S \qquad (11)$$

That is, the solution attractor is formed from the entire solution space S. The solution attractor A contains a unique minimal "convex" set for a unimodal instance, or h unique minimal "convex" sets $A_i (i = 1,2,\ldots,h)$ for a h-modal instance. Each A_i has a unique best tour s_i^* surrounded by a set of locally optimal tours. A best tour s_i^* in a solution attractor A_i satisfies $f(s_i^*) < f(s)$ for all $s \in A_i$ and $f(s_1^*) = f(s_2^*) = \cdots = f(s_h^*)$. Therefore, the "convexity" property of the solution attractor allows the propagation of the minimum property of s_i^* in the solution attractor A_i to the whole solution spacee S through the following conditions:

1) $\forall s \in N(s'), f(s') \leq f(s)$
2) $\forall s' \in A$ and $\forall s \in (S - A), f(s') \leq f(s)$
3) $\forall s' \in A, f(s_i^*) < f(s'), i = 1,2,\ldots h$

4) $f(s_1^*) = f(s_2^*) = \cdots = f(s_h^*)$
5) $\min_{s \in A} f(s) = \min_{s \in S} f(s)$

The global convergence and deterministic property of the search trajectories in the local search process make the ABSS always find the same set of globally optimal tours. We conducted several experiments to confirm this argument empirically. In our experiments, for a given TSP instance, the ABSS performed the same search process on the instance sevaral times, each time using a differetn set of k search trajectories. The search system outputed the same set of the best tours in all trials.

Table 1 shows the results of two experiments. One experiment generated $n = 1000$ instance Q_{1000}, the other generated $n = 10000$ instance Q_{10000}. We conducted 10 trials on each of the instances respectively. In each trial, the ABSS generated $K = 6n$ search trajectories. Each search trajectory stopped when no improvement was made during 10000 iterations for Q_{1000} and 100000 iterations for Q_{10000}. The matrix E stored the edge configurations of the K final tours and then was completely searched by the depth-first tree search process. Table 1 lists the number of tours in the constructed solution attractor A, the cost range of these tours, and the number of the best tours found in the constructed solution attractor. For instance, in trial 1 for Q_{1000}, the ABSS found 6475824 tours with the cost range [3241, 4136] in the constructed solution attractor. There was a single best tour in the solution attractor. The ABSS found the same best tour in all 10 trials. For the instance Q_{10000}, the ABSS found the same four best tours in all 10 trials. These four tours have the same cost value, but with different edge configurations. If any trial had generated a diffferent set of the best tours, we could immediately make a conclusion that the best tours in the constructed solution attractor may not be the globally optimal tours. From practical perspective, the fact that the same set of the best tours was detected in all trials provides an empirical evidence of the global optimality of these tours.

Table 1. Tours in solution attractor A for the 1000-nodes and 10000-nodes TSP instances

Trial Number	Number of Tours in A	Range of Tour Cost	Number of Best Tours in A
1000-nodes (Q_{1000})	(6000 search trajectories)		
1	6475824	[3241, 4136]	1
2	6509386	[3241, 3986]	1
3	6395678	[3241, 4027]	1
4	6477859	[3241, 4123]	1
5	6456239	[3241, 3980]	1
6	6457298	[3241, 3892]	1
7	6399867	[3241, 4025]	1
8	6423189	[3241, 3924]	1
9	6500086	[3241, 3948]	1
10	6423181	[3241, 3867]	1
10000-nodes (Q_{10000})	(60000 search trajectories)		
1	8645248	[69718, 87623]	4
2	8657129	[69718, 86453]	4
3	8603242	[69718, 86875]	4
4	8625449	[69718, 87053]	4
5	8621594	[69718, 87129]	4
6	8650429	[69718, 86978]	4
7	8624950	[69718, 86933]	4
8	8679949	[69718, 86984]	4
9	8679284	[69718, 87044]	4
10	8677249	[69718, 87127]	4

4.3. Computing Complexity of the ABSS

One of the requirements for a global optimization system is to find all optimal solutions within a *reasonable amount of computing time*. With current search technology, the TSP is an infeasible problem because it is not solvable in a reasonable amount of time. Faster computers will not help. A feasible search algorithm for the TSP is one that comes with a guarantee to find all best tours in time at most proportional to n^k for some power k. The

ABSS can guarantee to find all globally optimal tours for the TSP. Now the question is how efficient it is.

The focus of computational complexity theory is to analyze the intrinsic difficulty of the optimization problems and the asymptotic analysis of the algorithms to solve them. The complexity theory attempts to address this question: How efficient is an algorithm for a particular optimization problem, as the number of variables gets large? Most combinatorial problems including the TSP are known to be NP-hard. The problems in NP-hard are said to be intractable because these problems have no asymptotically efficient algorithm, even the seemingly "limitless" increase of computational power will not resolve their genuine intractability (Fortnow, 2013; Papadimitriou and Steiglitz, 1977; Vavasis, 1995). The real question in the computational complexity is about how fast we can search through a huge number of possibilities in a solution space. For the TSP, the computational complexity is associated with the combinatorial explosion of potential solutions in the solution space. When a TSP instance is large, the number of possible tours in the solution space is so large as to forbid an exhaustive search for the optimal tours. Do we need to explore all the possibilities to find the optimal solutions? The novel perspective of solution attractor in a local search system for the TSP gives us an opportunity to overcome combinatorial complexity. Local search trajectories can greatly reduce the search space for the exhaustive search, which means that the computational difficulty of the TSP can be dramatically reduced. The solution attractor shows us where we can find the globally optimal solution in the solution space. If we concentrate the exhaustive search effort only in this small region, rather than the entire solution space, the number of possibilities in search space is no longer prohibitive. The ABSS uses the following three principles to reduce computational complexity for the TSP:

1. Search trajectories in a local search system can quickly converge to the solution attractor that contains the globally optimal solutions;
2. The edge matrix E can transform the local search behavior of individual search trajectories into the global search behavior of the search system;

3. The solution atractor is small enough that can be quickly searched by an exhaustive search process.

The core idea of the ABSS is that, if we have to use exhaustive search to confirm the globally optimal points, we should first find a way to quickly reduce the effective search space for the exhaustive search. The local search phase in the ABSS can quickly prune out large number of edges that cannot possibly be included in any of the globally optimal tours. When the first edge is discarded by all local search trajectories, $(n-2)!$ tours that go through the edge are removed from the search space for the exhaustive search phase. Each time when an edge is removed, large number of branches are cut off from the search tree. In this way, the local search trajectories quickly reduce the number of combinatorial branching possibilities in the search space for the exhaustive search. The local search phase is for fast convergence and can be done in $O(n^2)$. Although the complexity of finding a true locally optimal tour is still open, and we even do not know any nontrivial upper bounds on the number of itertions that may be needed to reach local optimality (Chandra et al., 1999; Fischer 1995). In practice, we are rarely able to find perfect locally optimal tour because we simply do not allow the local search process to run enough long time. Usually we let a local search process run a predefined number of iterations, accept whatever tour it generates and treat it as a locally optimal tour. Therefore, the size of the constructed solution attractor depends not only on the problem structure and the neighborhood function, but also on the amount of search time invested in the local search process. If we increase local search time, we will construct a smaller and stronger solution attractor. Decades of empirical evidence and practical research have found that local search heuristics converge very quickly, within low-order polynomial time (Aarts and Lenstra, 2003; Applegate et al., 2006; Fischer, 1995; Grover, 1992).

The intrinsic difficulty of the TSP is that the solution space increases exponentially as the problem size increases, which makes the exhaustive search infeasible. An algorithm is said to be polynomial time if the number of iterations is bounded by a polynomial in the size of the problem instance. The local search phase in the ABSS reduces the search space for the exhaustive search to a solution attractor. Now an essential question is naturally raised about the size of the constructed solution attractor: What is the relationship between the size of the constructed solution attractor and the size of the problem instance? Unfortunately, there is no theoretical or analytical tool available in the literature that can be used to answer this question. We have to depend on empirical results to lend some insights. We conducted several experiments to observe the relationship between the size of the constructed solution attractor and the instance size. Figure 10-12 show the results of one of our experiments. All other similar experiments reveal the same pattern. In this experiment, we generated 10 unimodal TSP instances in the size from 1000 to 10000 nodes with 1000-node increment. For each instance, the search system generated $K = 6n$ local search trajectories. We first let each search trajectory stop when no tour improvement was made during 10000 iterations regardless of the size of the instance (named "fixed search time"). Then we did the same search procedures on these instances again. This time we made each search trajectory stop when no improvement was made during $10n$ iterations (named "varied search time 1") and $100n$ iterations (named "varied search time 2"), respectively. Figure 10 shows the number of edge discarded by the search trajectories at the end of the search, Figure 11 shows the number of tours in the constructed solution attractor for each instance, and Figure 12 shows the effective branching factors in the exhaustive search phase.

In Figure 10, we can see that the search trajectories can quickly converge to a small set of edges. In the fixed-search-time case, about 60% of edges are discarded by search trajectories for the 1000-node instance, but this percentage decreases as instance size increases. For the 10000-node instance, only about 46% of edges are discarded. However, if we increase the local search time linearly when the instance size increases, we can keep the same percentage of discarded-edge for all instance sizes. In the varied-search-time-1 case, about 60% of edges are abandoned for all different instances sizes. In the varied-search-time-2 case, this percentage increases to 68% for all instances. Higher percentage of abandoned edges manes that majority of branches are removed from the search tree.

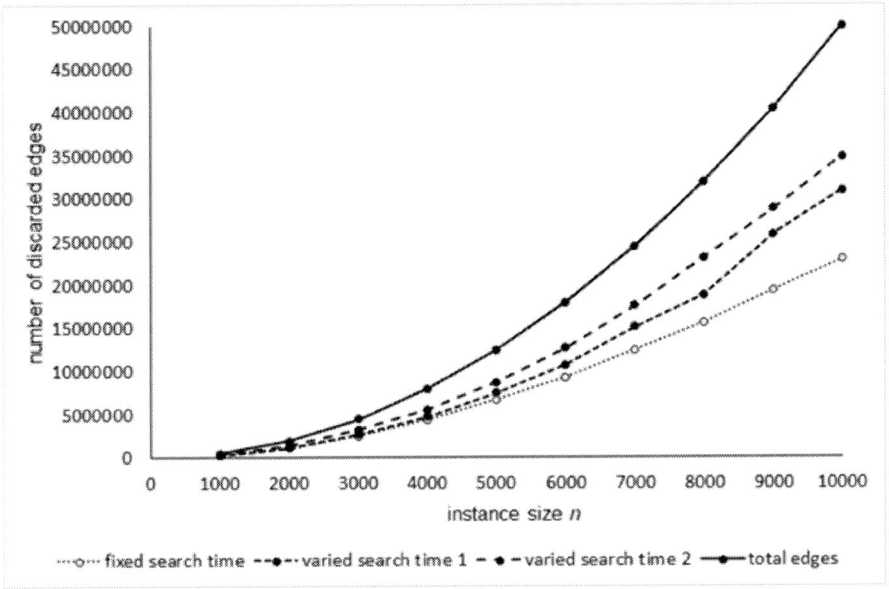

Figure 10. The number of discarded edges at the end of local search process.

Figure 11 shows the number of tours existing in the constructed solution attractor for these instances. All curves appear to be linear relationship between the size of constructed solution attractor and the size of the problem instance, and the varied-search-time curves have much flatter slope because longer local search time makes a smaller constructed solution attractor. Figure 10 & 11 indicate that the local search trajectories can effectively and efficiently reduce the search space for exhaustive search, and the size of the solution attractor may increase linearly as the size of the problem increases. Therefore, the local search phase in the ABSS is an asymptotically efficient search algorithm that produces an extremely small search space for further search in the exhaustive search phase.

The complete exploration of the solution attractor is delegated to the exhaustive search phase, which may still need to examine tens or hundreds of millions of tours but nothing a computer processor cannot handle, as opposed to the huge number of total possibilities in the solution space. The exhaustive search phase could find the exact globally optimal tours for the problem instance after a limited number of search steps.

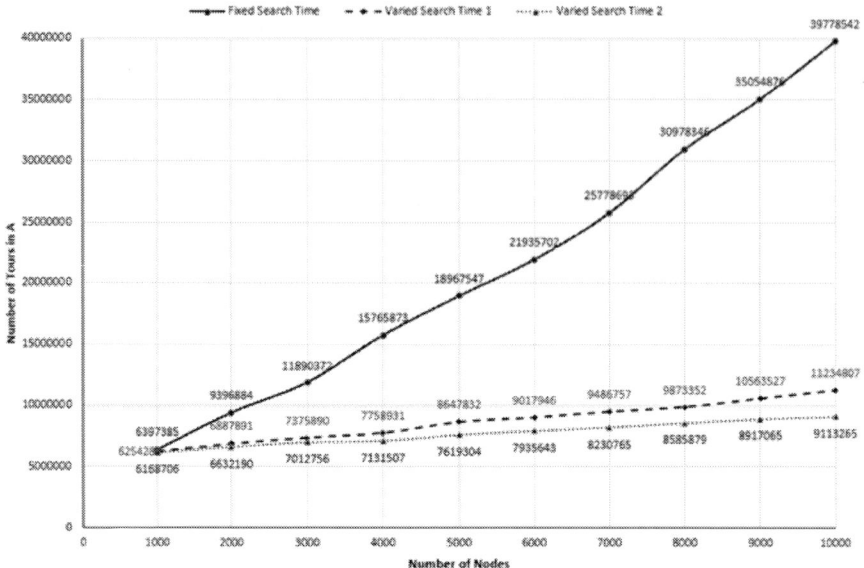

Figure 11. The relationship between the size of the constructed solution attractor and the size of the problem.

The exhaustive search phase can use any enumerative technique. However, the edge configuration of the matrix E can be easily searched by the depth-first search-tree algorithm. One of the advantages of depth-first tree search is that it requires less memory since only the nodes on the current path are stored. When using search-tree algorithm, we usually use branching factor, average branching factor, or effective branching factor to measure the computing complexity of the algorithm (Baudet, 1978; Edlkamp and Korf, 1998; Korf, 1985; Pearl, 1982). In a tree data structure, the branching factor is the number of successors generated by a given node. If this value is not uniform, an average branching factor can be calculated. An effective branching factor b^* is the number of sucessors generated by a typical node for a given search-tree problem. We use the following definition to calculate effective branching factor b^* in the exhaustive search phase:

$$N = b^* + (b^*)^2 + \cdots + (b^*)^n \qquad (12)$$

where n is the size of the TSP instancee, representing the depth of the tree;

N is total number of nodes generated in the tree from the origin node. In our experiments, the search-tree process always starts from node 1 (the first row of the matrix E). N is total number of nodes that are processed to construct all valid tours and incomplete (therefor abandoned) tours in the matrix E. N does not count the node 1 (the origin node), but includes the node 1 as the end node of a valid tour.

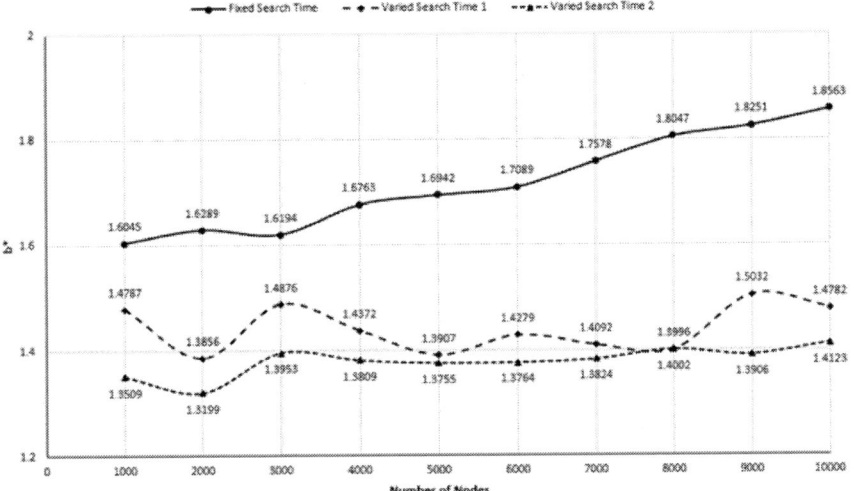

Figure 12. The b^* values for different problem size n.

Figure 12 shows the effective branching factor s^* for the instances reported in Figure 10. The result indicates that the edge configuration of the solution attractor presents a tree with extremely sparse branches, and the degree of sparseness does not change as the problem size increases if we properly increase local search time for larger instances. The search time in the exhaustive search phase probably is in $O(n^2)$ since the size of a constructed solution attractor might be linearly increased with the problem size n and the number of edges in the matrix E is algebraically increased with the problem size. Our experiments showed that the local search trajectories can significantly reduce the search space for the exhaustive search and the ABSS can effectively reduce computational complexity of the TSP.

Therefore, the ABSS is a simple algorithm that increases in difficulty polynomially with the size of the TSP. In the ABSS, the objective pursued by the local search phase is "collecting all most promising tours in the solution space as quickly as possible." It can provide an answer to the question: "In which small region of the solution space is the optimal solution located?" in time of $O(n^2)$. The objective of the exhaustive search phase is "identifying the exact best tour among the most promising tours". It can provide an answer to the question: "In this small region, which solution is the best one?" in time of $O(n^2)$. All together, the ABSS can answer the question "Is this tour the optimal tour in the solution space?" in time of $O(n^2)$. In summary, the ABSS probably is in computing complexity of $O(n^2)$ and memory space requirement of $O(n^2)$. This suggests that the TSP might not be as complex as we might have expected.

CONCLUSION

Advances in computational techniques on the determination of the global optimum of an optimization problem can have great impact in many scientific and engineering fields. Although both the TSP and local search algorithms have huge literature, and numerous experts have made huge advance in the TSP research, the fundamental question of the TSP remains essentially open. The study of local search for the TSP continues to be an important research problem in computer science, computational mathematics and combinatorial optimization. So far, there exists no efficient algorithm to truly solve the TSP. Only the exhaustive search algorithm can provide an exact solution. Modern computers can greatly speed up the search, but the extremely large solution spaces would still require geologic search time to find the exact optimal solution on the fastest machines imaginable. Huge number of possibilities in the solution space does not mean that you need to explore them all. Most people believe $P \neq NP$ because we have made little fundamental progress in the area of exhaustive search. The nature of P vs. NP problem is about how much we can make improvement upon exhaustive search, in general, when solving combinatorial problems. A new point of view is needed to improve our ability to tackle these difficulty problems. One possible way to solve such a problem practically would be to find a method that can effectively reduce the search space for exhaustive search. The concept of solution attractor in local search systems may change the way we think about both local search and exhaustive search. In fact, the goal of any optimization problem requires us to design a search system that is both effective and efficient. Heuristic local search is an efficient search system in which the search trajectories converge quickly to a small region in the solution space; while the exhaustive search is an effective search system that can find the exact optimal solutions. The key is how we combines these two systems into one system beautifully to conquer the fundamental issues of the hard optimization problems. In our case, the edge matrix E (the problem-specific data structure) plays a critical role to transform local search to global search and reduce the search space for exhaustive search.

REFERENCES

Aart, E. and Lenstra, J. K. (2003) *Local Search in Combinatorial Optimization*. Princeton: Princeton University Press.

Alligood, K. T., Sauer, T. D. and York, J. A. (1997) *Chaos: Introduction to Dynamical System*. New York: Springer-Verlag.

Applegate, D. L., Bixby, R. E., Chaátal, V. and Cook, W. J. (2006) *The Traveling Salesman Problem: A Computational Study*. Princeton: Princeton University Press.

Auslander, J., Bhatia, N. P., and Seibert, P. (1964) Attractors in dynamical systems. *NASA Technical Report NASA-CR-59858*.

Baudet, G. M. (1978) On the branching factor of the alpha-beta pruning algorithm. *Artificial Intelligence*. 10(23), 173-199.

Brin, M. and Stuck, G. (2016) *Introduction to Dynamical Systems*. Cambridge: Cambridge University Press.

Brown, R. (2018) *A Modern Introduction to Dynamical Systems*. New York: Oxford University Press.

Buescu, J. (1991) *Exotic Attractors*. Basel: Birkhäuser.

Chandra, B., Karloff, H. J. and Tovey, C. A. (1999) New results on the old-opt algorithm for the traveling salesman problem. *SIAM Journal on Computing*. 28(6), 1998-2029.

Collet, P. and Eckmann, J. P. (1980) *Iterated Maps on the Interval as Dynamical System*. Boston: Birkhäuser.

Dénes, A. and Makey, G. (2011) Attractors and basis of dynamical systems. *Electronic Journal of Qualitative Theory of Differential Equations*. 20(20), 1-11.

Edelkamp, S. and Korf, R. E. (1998) The branching factor of regular search space. *Proceedings of the 15th National Conference on Artificial Intelligence*. pp. 229-304.

Fischer, S. T. (1995) A note on the complexity of local search problems. *Information Processing Letters*. 53(2), 69-75.

Fogedby, H. (1992) On the phase space approach to complexity. *Journal of Statistical Physics*. 69, 411-425.

Fortnow, L. (2013) *The Golden Ticket – P, NP, and the Search for the Impossible*. Princeton University Press.

Gagniuc, P. A. (2017) *Markov Chains: From Theory to Implementation and Experimentation*. Hoboken: John Wiley & Sons.

Gomey, J. (2019) Stochastic global optimization algorithms: a systematic formal approach. *Information Science*. 472, 53-76.

Grove, L. K. (1992) Local search and the local structure of NP-complete problems. *Operations Research Letters*. 12(4), 235-243.

Hastings, W. K. (1970) Monte Carlo sampling methods using Marko chains and their applications. *Biametnika*. 57(1), 97-109.

Horst, R. and Pardalos, P. M. (1995) *Handbook of Global Optimization: Nonconvex Optimization and Its Application*. New York: Springer.

Johnson, D. (1990) Local optimization and the traveling salesman problem. *Automata, Languages and Programming: Proceedings of the 17th International Colloquium*. New York: Springe, 446-461.

Kolokoltsov, V. (2010) *Nonlinear Markov Process*. Cambridge: Cambridge University Press.

Korf, R. E. (1985) Depth-first iterative deepening: an optimal admissible tree search. *Artificial Intelligence*. 27, 97-109.

Korte, B. and Vygen, J. (2007) *Combinatorial Optimization: Theory and Algorithms*. New York: Springer.

Kroese, D. P., Taimre, T. and Botev, Z. I. (2011) *Handbook of Monte Carlo Methods*. New York: John Wiley & Sons.

Li, W., (2005). Dynamics of local search trajectory in traveling salesman problem. *Journal of Heuristics*. 11, 507-524.

Li, W. and Feng, M. (2013) Solution attractor of local search in traveling salesman problem: concepts, construction and application. *International Journal of Metaheuristics*. 2(3), 201-233.

Li, W. and Li, X. (2019) Solution attractor of local search in traveling salesman problem: computational study. *International Journal of Metaheuristics*. 7(2), 93-126.

Lin, S. (1965) Computer solution of the traveling salesman problem. *Bell System Technical Journal*. 44, 2245-2269.

Mahfoud, S. W. (1995) *Niching methods for genetic algorithms*. Ph.D Dissertation. University Illinois Urbana-Champaign.

Masel, J. (2011) Genetic drift. *Current Biology*. 21(20), R837-R838.

Meyn, S. and Tweedie, R. (2009) *Markov Chains and Stochastic Stability*. New York: Cambridge University Press.

Michalewicz, Z. and Fogel, D. B. (2002) *How to Solve it : Modern Heuristics*. Berlin: Springer.

Milnor, J. (1985) On the concept of attractor. *Communications in Mathematical Physics*, 99(2), 177-195.

Milnor, J. (2010) *Collected Papers of John Milnor VI : Dynamical Systems (1953-2000)*. American Mathematical Society.

Papadimitriou, C. H. and Steiglitz, K. (1977) On the complexity of local search for the traveling salesman problem. *SIAM Journal on Computing*. 6(1), 76-83.

Papadimitriou, C. H. and Steiglitz, K. (1998) *Combinatorial Optimization : Algorithms and Complexity*. New York: Dover Publications.

Pardalos, P., Van Thoai, N. and Horst, R. (2000) *Introduction to Global Optimization: Nonconvex Optimization and Its Applications*. New York : Springer.

Pearl, J. (1982) The solution for the branching factor of the alpha-beta pruning algorithm and its optimality. *Communication of the ACM*. 25(8), 559-564.

Rayward-Smith, V., Osman, I. H., Reeves, C. R. and Smith G. D. (1996) *Modern Heuristic Search Methods*. New York: Wiley.

Reeves, C. R. (1993) *Modern Heuristic Techniques for Combinatorial Problems*. New York: Wiley.

Rego, C., Gamboa, D., Glover, F. and Osterman, C. (2011) Traveling salesman problem heuristics: leading methods, implementations and latest advances. *European Journal of Operational Research*. 211(3), 427-441.

Ruelle, D. (1981) Small random perturbations of dynamical systems and the definition of attractor. *Communications in Mathematical Physics*. 82(1), 137-151.

Sahni, S. and Gonzales, T. (1976) P-complete approximation problem. *Journal of the ACM*. 23, 555-565.

Savage, S. L. (1976) Some theoretical implications of local optimality. *Mathematical Programming*. 10, 354-366.

Shammon, C. E. (1948) A mathematical theory of communication. *Bell System Technical Journal*. 27: 379-423, 623-656.

Sourlas, N. (1986) Statistical mechanics and the traveling salesman problem. *Europhysics Letters*. 2(12), 919-923.

Vavasis, S. A. (1995) Complexity issues in global optimization : a survey. *Handbook of Global Optimization : Nonconvex Optimization and Its Application*. New York: Springer, 27-41.

Zhigliavsky, A. and Žillinakas, A. (2008) *Stochastic Global Optimization*. New York: Springer.

In: The Fundamentals of Search Algorithms ISBN: 978-1-53619-007-6
Editor: Robert A. Bohm © 2021 Nova Science Publishers, Inc.

Chapter 2

BIOMETRIC DATA SEARCH ALGORITHM

*Stella Metodieva Vetova**, *PhD*
Department of Informatics, Technical University of Sofia,
Sofia, Bulgaria

ABSTRACT

The purpose of the outlined content is to show the practical application of CBIR technology in the security and protection of personal data, access to classified documents and objects, identification of illegal attacks that are part of the social life of the present and future of mankind. It involves comparison of the disadvantages and advantages of content-based image retrieval techniques with the use of local and global features in biometric data. The main idea is to offer the optimal application for security and protection. In pursuit of this goal, the following tasks are envisaged:

1. Design of algorithms with the use of local characteristics and
2. different similarity distance measures;
3. Design of algorithms with the use of global characteristics using
4. different similarity distance measures;
5. Conducting experimental studies on the algorithms of items 1 and 2;
6. Comparative analysis of the data obtained in item 3.

---

* Corresponding Author's Email: vetova.bas@gmail.com.

Considering this, two algorithms with different similarity distance measures were developed using the two techniques, and in this case Hausdorff distance and Euclidean distance were chosen as distance metrics. For the representation of a two-dimensional signal through its feature vectors, the decomposition of the signal to approximate and detailed coefficients by the 2D Dual - Tree Complex Wavelet Transform was used. The concept of this process is described visually by graphic illustrations, diagrams, images and analyzes. Also, a description of the research methodology, test databases and used software is included. To evaluate the effectiveness of the designed algorithms according to both techniques, test studies were conducted to evaluate the accuracy of the extracted result, the retrieval time, the number of retrieved images. The results are presented graphically, in tables and by image. At the end of the chapter, a comparative analysis between the algorithms is made on the base of the results obtained, which shows the advantages and disadvantages of both techniques and their application in the recognition of biometric data.

Keywords: CBIR, biometrical data, DT CWT, Hausdorff distance, Euclidean distance, local feature vectors, global feature vectors, wavelet transform, wavelet coefficients

1. INTRODUCTION

The idea of content-based image retrieval (CBIR) is based on the successful automatic recognition of a query-image. To this end, feature vectors, also called descriptors, are used, and a similarity measure is used to determine similar images. The idea is to retrieve images according to the following requirements:

- maximum extraction rate;
- accuracy of the result obtained [1];
- minimum size of memory for data recording.

Technology was born in 1992 [2] as an alternative to Text-Based Image Retrieval (TBIR). It uses text keywords to form the query that describe the image for retrieval. This has three major drawbacks. First, a human factor is needed to introduce the textual description of images, which is expensive for technology. Second, the process of typing text descriptions requires time, which converts the image retrieval into a time-consuming process [3]. Third, the result obtained is directly dependent on human perception [4], which is leading in the process of description. By overcoming these drawbacks, content extraction proves its objectivity and efficiency, expressed in the higher speed and accuracy of the extracted result, as well as the lower cost of the technological process. This technology performs two main tasks. The first one is the image feature extraction, using a technique to generate a feature vector. This vector is the main storage medium for the image and presents it in the Image Data Base (IDB), requiring a smaller size of memory for recording data than storing it in graphical format. The second task is the definition of similarity (measurement), which determines the similarity between the query-image and each of the images contained in the image database using their feature vectors. The images with the highest rank of similarity (rank position in the series of displayed images; the highest rank is the first position) are displayed as the final result of the user search.

Due to its advantages, CBIR is widely used in a wide range of areas [2]:

- Face recognition and fingerprints in criminology;
- Biometrical data identification;
- Detection of defects in industrial automation;
- Tumor recognition, refinement of nuclear magnetic resonance (NMR) technology, and more in medical diagnostics;
- Remote monitoring systems such as GIS, satellite imaging systems, weather forecasting systems, etc.
- Database (DB) for trademarks, art galleries, museums, archaeological exhibitions;
- Architectural and engineering design;
- Creating maps of photographs in cartography;
- Recognition of targets in radar engineering;
- Fashion, graphic design, advertising, publishing.

A recent trend in the development of CBIR technology is the development of CBIR mobile technology applications [5].

The techniques used to extract feature vectors use low-level features such as color, texture, shape, and spatial layout, and a combination of these to achieve higher accuracy of the final result. They are grouped into two main categories: spatial and spectral.

One of the CBIR applications in the biometrical authentication branches is iris analysis. Iris is a circular eye structure characterized by color, texture, and shape. It has unique texture which does not change throughout a person's life and makes it suitable for authentication in such fields as criminology, immigration control, and national identification systems. Capturing the unique information in the iris texture structure of each of a person's eyes rises the accuracy of authentication and therefore security against threads. Moreover, it provides fast and accurate authentication with no ID card required and especially in the cases when combined with face, fingerprint, and palm print recognition.

The technology for capturing iris pattern requires illuminating the eyes with harmless infrared light so that eye details to be captured successfully nevertheless the iris color is light of dark.

The technology of recognition and retrieval of biometric data varies. One of the most used methodologies include segmentation [6, 7], normalization, feature vectors extraction and similarity or matching computation. Chitra and Amsavalli Subramanian [8] apply segmentation to separate the iris region by circular Hough transformation and encode it using logarithmic Gabor Wavelets. A similar technique was implemented by Vimala and Karthika Pragadeeswari [9]. After segmentation using the Sobel operator, they apply the Fast Fourier Transform for iris feature vector generation. On the other hand, in [10] authors present their own method for iris recognition, using Hough transform for segmentation, and for iris feature extraction - FLDA/PCA fusion iris extraction scheme. The encoding is accomplished by a 1D Log-Gabor filter. To detect shape, an invariant moment and the Canny edge filter are used [11].

A common measure of similarity is Hamming distance [8, 9, 10, 12].

Beside the technological solution for biometric data described above, another often used method is the one based on advanced training techniques. In [14] authors offer a deep learning framework, DeepIrisNet2, for generating binary iris codes. It ignores the need to normalize the iris and consists of spatial transformer layers to handle deformation. They also present a dual CNN iris segmentation pipeline for eye and pupil iris detection.

On the other hand, Wei Zhang et al. [15] offer four executable network circuits. Fully Dilated Convolution combining U-Net (FD-UNet) is a model designed by training and testing using the same datasets and demonstrating the best ability to recognize iris patterns. FD-UNet is designed on the base of dilated convolution to achieve higher results extracting more global features contributing to better image details processing.

In [16], authors propose a combined method consisting of multiclassification method using the Hadamard error correction output code and a convolutional neural network.

In addition to the wide used classification methods [17, 18, 19, 20] K-Means clustering is proposed. In order to improve its performance, it is combined with sensitive and dependent upon the initial centroid procedure.

This chapter presents a CBIR method for iris biometrical data using the Dual-Tree Complex Wavelet Transform for generating feature vectors and Hausdorff distance and Euclidean distance for similarity computation. Both techniques for feature extraction based on local and global features are applied and the results obtained compared and analyzed through a comparative analysis. The experimental work uses four algorithms using both feature extraction strategy and Hausdorff distance and Euclidean distance.

2. CONTENT-BASED IMAGE RETRIEVAL APPROACH FOR BIOMETRIC DATA ANALYSIS

The approach presented in this chapter is based on the idea of locally retrieving features and generating local feature vectors whose task is to extract more detailed information from images. Two algorithms have been developed - an algorithm using local features and Hausdorff distance (ALFH) and an algorithm using local features and Euclidean distance

(ALFE). They both use two-dimensional images represented in greyscale color space, allowing only one component to work. This has the effect of achieving greater speed than the technology of extracting feature vectors based on the three components that compose color digital images. In addition, color invariance of a small range of the same subjects recorded with slight differences in shooting conditions is ensured. The defining characteristics of the objects under study are their shape, texture and degree of contrast in the identification areas. Taken at different degrees of illumination by a sensor of different types and noise characteristics, the greyscale image is an invariant medium of object information compared to the color digital image.

In Figure 1 a flowchart of the CBIR algorithm for image retrieval using local feature vectors is presented. It describes the sequence of steps in implementing the proposed approach using Hausdorff distance or Euclidean distance to determine the degree of similarity between the query-image and those in the test database.

The proposed approach is implemented in four stages as follows: Image Submission, Image Preprocessing, Image Content Analysis, Similarity Measurement Analysis. The first step involves loading and reading the image (from the test database or query-image). The second stage includes Image scale and Image subdivision, which are accomplished in the following order, respectively:

- the source image (I), is resized to the size of *MxN* for *M = N = 256* where M - number of rows, N - number of columns;
- the image (I'') is divided into $n \times n$ ($n = 8$) non-overlapping subimages ($I_1'' \div I_{64}''$).

Image Content Analysis covers the steps of analyzing image content through Dual-Tree Complex Wavelet Transform (DT CWT) [21, 22, 23, 24, 25, 26, 27, 28, 29] as follows:

- DT CWT is performed on each of the subimages $(I_1^{"} \div I_{64}^{"})$ at level $l = 4$ to decompose them and generate their lowpass and bandpass components;
- the extracted wavelet features form the image feature vectors which are stored in a CBIR database.

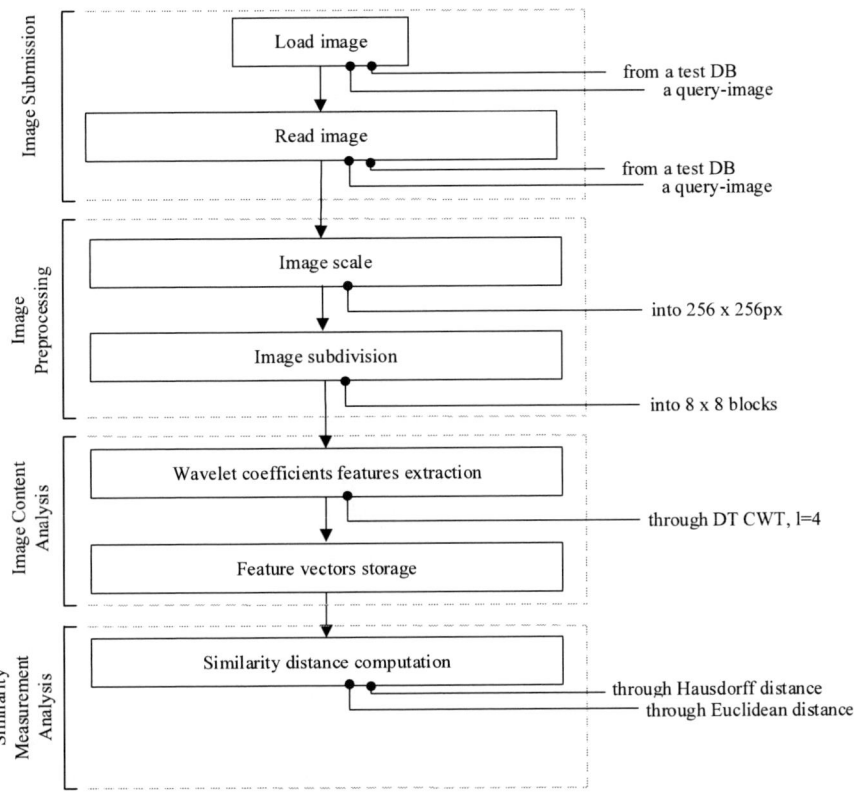

Figure 1. Flowchart of the CBIR algorithm for image retrieval using local feature vectors.

The last stage describes the step of similarity distance computation using two metrics for this purpose: Hausdorff distance and Euclidean distance.

On the other hand, in the research work on analyzing iris biometric data, the performance and results of two other algorithms are analyzed too – the algorithm using global features and Hausdorff distance (AGFH) and the algorithm using global features and Euclidean distance (AGFE). Their technological solution differs from that described of the algorithms using local characteristics by DT CWT performance on an entire image without image subdivision.

3. Image Test Database

CASIA-IrisV4 is an image test database containing six subsets of iris biometrical data as follows: CASIA-Iris-Interval, CASIA-Iris-Lamp, CASIA-Iris-Twins, CASIA-Iris-Distance, CASIA-Iris-Thousand, and CASIA-Iris-Syn. CASIA-IrisV4 consists of 54,601 iris images of 8 bit gray-level JPEG files. Some of the test images have been designed using infrared illumination while others have been synthesized. Portions of the research in this chapter use the CASIA-IrisV4 collected by the Chinese Academy of Sciences' Institute of Automation (CASIA) [30]. For the presented experimental tests and results in the chapter CASIA-Iris-Lamp subset is used, consisting of 8220 JPEG files of 8 bit gray-level color space.

4. Hardware Configuration

The design and test experiments of the developed approach and algorithms is implemented through Matlab software for mathematical and engineering computations. It is installed on a computer configuration with a 64-bit operating system (OS) and hardware specification: Intel (R) Core (TM) i5-2400, 3.10 GHz.

5. RESEARCH METHODOLOGY

5.1. Dual Tree Complex Wavelet Transform Decomposition Level Choice for Feature Vectors Extraction

Designing an algorithm that meets the requirements for minimum memory for data recording, high accuracy and image retrieval speed requires working with feature vectors whose coefficients are information carriers and are minimal in quantity. To this end, a study for image decomposition level using DT CWT was conducted for each decomposition level l ($l = 1, 2, 3, 4$) [31].

5.2. CBIR Retrieval Time Evaluation

The test image I_{n-1} from the test database is submitted as a query-image I_q. After the feature vector generation, a process of computing similarity using the Hausdorff distance $H(A,B)$ or the Euclidean distance $d_{Euc}(A,B)$ starts. The number of accomplished comparisons corresponds to the number of feature vectors - one representing the corresponding image I_{n-1} in the test image database. When this procedure completes, the time duration is measured in seconds (ExT_p). On the basis of all the experiments performed (p) (15 for each of the studied algorithms), the arithmetic mean time (ExT_{avr}) for image retrieval according to formula (1) is computed:

$$ExT_{avr} = \frac{\sum_{p=1}^{15} ExT_p}{15} \tag{1}$$

The described methodology is applied as for conducting experimental studies based on both local and global features.

5.3. Efficiency Evaluation without Rank

The classic method of determining performance (excluding rank) includes the Precision and Recall metrics. They are defined by expressions (2) and (3) respectively:

$$Precision = \frac{|relevant \cap retrieved|}{|retrieved|}, (\%) \qquad (2)$$

$$Recall = \frac{|relevant \cap retrieved|}{|relevant|}, (\%) \qquad (3)$$

The first of the metrics (Precision) is the ratio of the number of relevant retrieved images and the number of all retrieved images. At Precision = 100%, all retrieved images are considered to be relevant.

The second metric (Recall) is the ratio of the number of relevant retrieved images and the number of all relevant images in the test image database. Achieving Recall = 100% is associated with retrieving the entire test image database, resulting in reduced precision metric values that tend to zero. To evaluate the effectiveness of the entire test image database, fifteen experimental studies are conducted by submitting a query-image from each of the fifteen tested categories. The results obtained are analyzed with respect to Precision and Recall according to formulas (2) and (3).

6. AGFH Efficiency Evaluation without Rank

In order to analyze and evaluate AGFH, experimental studies were performed according to the described methodology. The choice of query-image is based on the requirement of minimum ten extracted images as retrieved result. Table 1 lists the results obtained for the fifteen experiments performed.

Table 1. Experimental results for AGFH efficiency evaluation

CASIA-IRIS-LAMP GROUPS	Relevant Retrieved Images	Irrelevant Retrieved Images	Retrieved Images	Precision (%)	Recall (%)
Group A	5	5	10	50	25
Group B	7	3	10	70	35
Group C	7	3	10	70	35
Group D	10	0	10	100	50
Group F	9	1	10	90	45
Group G	10	0	10	100	50
Group H	10	0	10	100	50
Group J	10	0	10	100	50
Group K	10	0	10	100	50
Group M	7	3	10	70	35
Group N	10	0	10	100	50
Group P	7	3	10	70	35
Group Q	10	0	10	100	50
Group S	10	0	10	100	50
Group Z	10	0	10	100	50

In Figure 2 and Figure 3 the retrieved result of the accomplished tests using Group A and Group M is presented. Comparison by group affiliation and shape recognition based on the data in Table 1 between the studied groups shows that the highest scores are characterized by nine of them: Group D, Group G, Group H, Group J, Group K, Group N, Group Q, Group S, Group Z. A similar result distinguishes Group F, and for the other four, Group B, Group C, Group M, Group P, the algorithm shows satisfactory results. For Group A, the algorithm demonstrates low recognition ability. The algorithm is distinguished by its ability to produce the best results in recognizing images in the center of which homogeneous regions are located, regardless of whether the intensity is high or low according to the query-image. On retrieval, the algorithm provides images with transitions expressed both in sharp and sharp jumps and in shimmering tones or transitions from areas of lower intensity to those with higher intensity and outstanding texture.

The algorithm extracts images relevant to the query-image by recognizing centrally located objects of the same or similar size with a low intensity pixel concentration. The algorithm also recognizes areas radially located (iris) on the central object (pupil), which are characterized by both the same pixel intensity as requested and varying in lower or higher intensity values. It is characterized by the ability to retrieve images with a distinct dark circle around the iris or those with a less outlined circle unlike the request (Figure 2). In addition, the algorithm successfully recognizes transitions expressed in sharp and sharp jumps from higher to lower intensities and vice versa, making it suitable for texture recognition, in this case radial furrows in the structure of the iris, represented in greyscale color space. When submitted with radially furrowed radial furrows, AGFH skips images with similar texture and crypts, as well as variations of pupillary area, ciliary area, collarette. In addition, AGFH demonstrates sensitivity with respect to the distance between the upper and lower eyelids, extracting images at the same or close distance from that of the query-image (Figure 2, Figure 3).

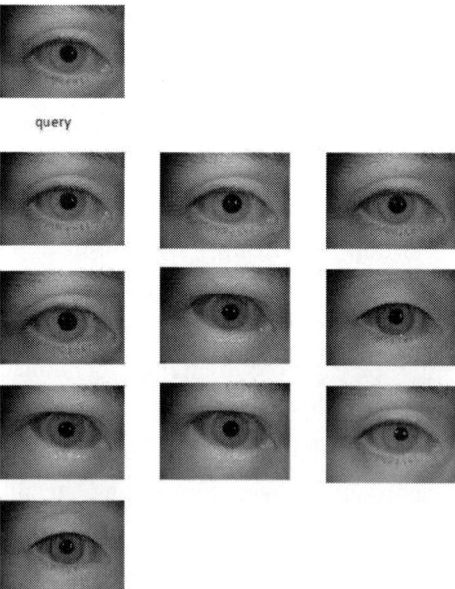

Figure 2. Extracted images upon request from Group A for AGFH.

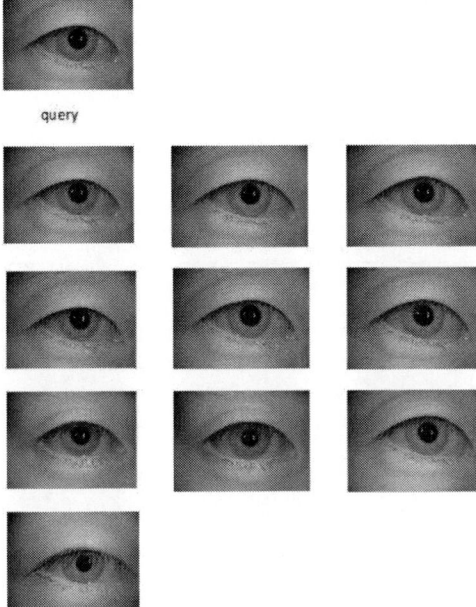

Figure 3. Extracted images upon request from Group M for AGFH.

7. AGFE Efficiency Evaluation without Rank

Studies have been conducted to determine the effectiveness of AGFE. To this end, the methodology described above is applied. The choice of query-image is based on the requirement of minimum ten extracted images as retrieved result. The results obtained for the studied image groups are shown in Table 2.

In Figure 4 and Figure 5 the retrieved result of the accomplished tests using Group A and Group M is presented. The data in Table 2 show that the best recognition is observed in the categories Group C, Group D, Group J, Group K, Group Z, where the group affiliation and form performance is 100%. From the cited categories for Group F and Group G, on these indicators the percentage of accuracy of the relevant images extracted from the total number of images extracted is 90%. For the other categories tested, accuracy values range between 80% and 50%.

Table 2. Experimental results for AGFE efficiency evaluation

CASIA-IRIS-LAMP GROUPS	Relevant Retrieved Images	Irrelevant Retrieved Images	Retrieved Images	Precision (%)	Recall (%)
Group A	5	5	10	50	25
Group B	7	3	10	70	35
Group C	10	0	10	100	50
Group D	10	0	10	100	50
Group F	9	1	10	90	45
Group G	9	1	10	90	45
Group H	8	2	10	80	40
Group J	10	0	10	100	50
Group K	10	0	10	100	50
Group M	7	3	10	70	35
Group N	8	2	10	80	40
Group P	8	2	10	80	40
Group Q	8	2	10	80	40
Group S	8	2	10	80	40
Group Z	10	0	10	100	50

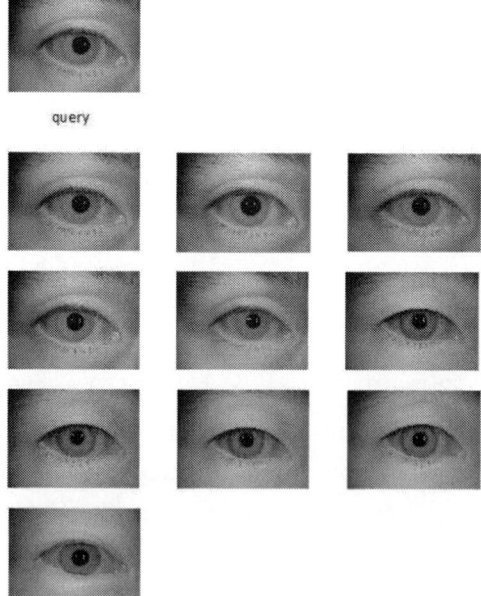

Figure 4. Extracted images upon request from Group A for AGFE.

Biometric Data Search Algorithm 59

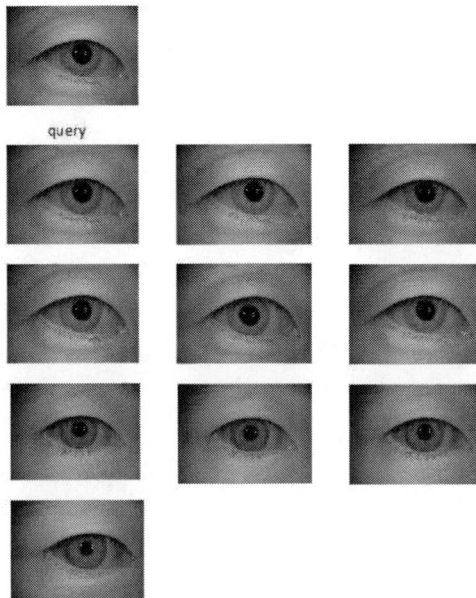

Figure 5. Extracted images upon request from Group M for AGFE.

The algorithm is distinguished by its ability to produce the best results in recognizing sharp jumps between regions of different intensities, shapes and textures, and the transition can be between different brightness regions with a transition from higher to lower intensity and vice versa. The location of the objects is central in accordance to the query-image. Higher or lower intensity pixel concentration may also be located in the center of the image corresponding to the query. The algorithm recognizes even the smooth transitions between entire areas with an intensity similar to its distribution in the query-image itself.

AGFE demonstrates the ability to recognize radially located areas representing the structure of the iris around the central object (pupil). The intensity of the corresponding pixels is characterized by both the same value according to the query and varying in lower or higher values. AGFE also has the ability to retrieve images with a distinct dark circle around the iris or those in which this detail matches the query. Successful recognition of transitions, expressed in sharp jumps from higher to lower intensity, and vice versa contributes to a high degree of recognition of the structure of the iris, incl. radial furrows. For the cases of a query-images containing radial furrows, AGFE retrieves 98% of images with a similar iris structure and 2% with such containing crypts. The result is similar when evaluating pupillary area, ciliary area, collarette. The algorithm studied is sensitive to the distance between the upper and lower eyelids, extracting images at the same or close distance from that of the query (Figure 4, Figure 5).

8. ALFH Efficiency Evaluation without Rank

The purpose of the research is to evaluate and analyze the effectiveness of the designed CBIR approach and algorithm. They were carried out according to the methodology described. The selection of a query is consistent with the generation of a minimum of ten number of images, and the resulting table results are based on the first 10 retrieved images from the query. The results obtained are presented in Table 3 for all fifteen studied categories of iris images.

Table 3. Experimental results for ALFH efficiency evaluation

CASIA-IRIS-LAMP GROUPS	Relevant Retrieved Images	Irrelevant Retrieved Images	Retrieved Images	Precision (%)	Recall (%)
Group A	8	2	10	80	40
Group B	7	3	10	70	35
Group C	10	0	10	100	50
Group D	10	0	10	100	50
Group F	7	3	10	70	35
Group G	10	0	10	100	50
Group H	10	0	10	100	50
Group J	10	0	10	100	50
Group K	10	0	10	100	50
Group M	10	0	10	100	50
Group N	8	2	10	80	40
Group P	10	0	10	100	50
Group Q	10	0	10	100	50
Group S	8	2	10	80	40
Group Z	9	1	10	90	45

The high precision values prove the efficiency of the algorithm in accordance with the retrieved images.

Based on a comparative analysis of the data from Table 3 by group affiliation and pattern recognition among the highest-performing groups studied, nine of them are characterized: Group C, Group D, Group G, Group H, Group J, Group K, Group M, Group P, Group Q. With a close result, Group Z is distinguished, and for the other two, Group B and Group F, the algorithm shows satisfactory retrieval results.

In Figure 6 and Figure 7, the presented images are retrieved on request, for the categories Group A and Group M. To evaluate the efficiency of the algorithm without rank, results are analyzed by group affiliation, shape recognition and texture required for the correct recognition of iris biometric data.

The algorithm demonstrates good results in the recognition of forms, which is comes as a result of the similarity distance computation by Hausdorff distance. ALFH also recognizes texture in the images. This is clearly evident when recognizing the radial furrows of the iris in the figures shown.

The algorithm is distinguished by its ability to perform best in image recognition where a concentration of pixels of close or equal intensity in central areas such as the pupil and iris of the eye are present, leading to increased recognition by group affiliation. Correctly extracted images from image database have similar details to the query-image.

The algorithm recognizes images both in smooth transitions between higher and lower intensity areas and in sharp transitions between areas of small size, equal complexity, repeatability and contrast, the latter affecting texture definition and recognition. Recognized objects have the orientation and location matching the query-image. Images of the same category that have a near or higher average brightness than the query are correctly extracted from the image database, which indicates that the algorithm has low sensitivity to changing this parameter.

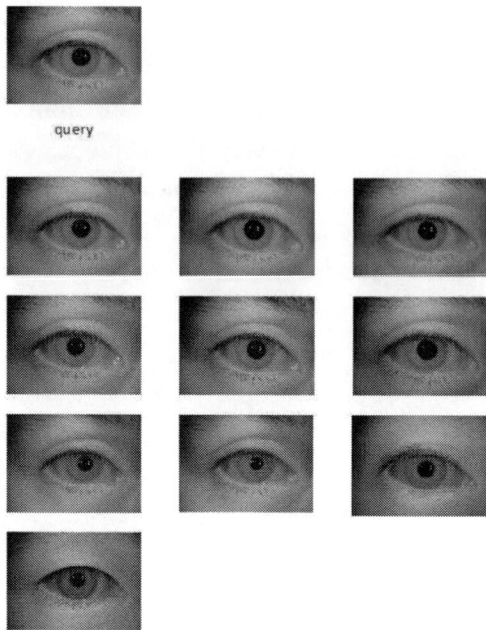

Figure 6. Extracted images upon request from Group A for ALFH.

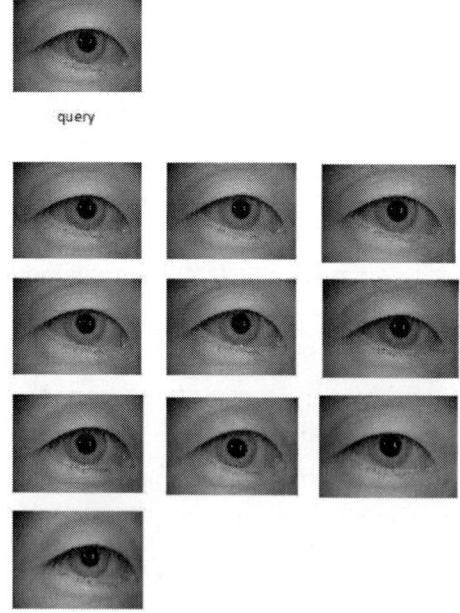

Figure 7. Extracted images upon request from Group M for ALFH.

ALFH also recognizes areas radially located (iris) at the central object (pupil) that are characterized by the same pixel intensities according to the query as well as by close intensity values. When submitted with radial furrows, ALFH retrieves in the first ten positions of the extracted images those with the same or similar texture, pupillary area, ciliary area, collarette (Figure 6). In addition, ALFH demonstrates the ability to retrieve images belonging to the same group, characterized by different pupil size (Figure 6). The algorithm is also distinguished by a slight sensitivity of 2% with respect to the distance between the upper and lower eyelids (Figure 6, Figure 7).

9. ALFE Efficiency Evaluation without Rank

Experimental studies have been conducted to evaluate the effectiveness of ALFE. They are performed according to the methodology described above. From the research work done on fifteen categories of images, the results obtained for two of them are selected and presented. The choice of query-image is based on the requirement of minimum ten retrieved images. The result is listed in Table 4.

The retrieved result for Group A and Group M is presented in Figure 8 and Figure 9. The evaluation of the effectiveness of the algorithm without rank is held on the basis of two indicators: group affiliation and shape recognition. In the first one, the algorithm presents the best results for the Group C, Group G, Group H, Group J, Group K, Group Q. For the six categories, the algorithm shows 100% of shape recognition.

The algorithm is distinguished by its ability to produce the best results in recognizing images containing centrally located objects, such as the pupil and iris when recognizing biometric data, where concentration of pixels with close or identical intensity and shape values is observed. It skips images belonging to groups different from that of the query-image, that is, three of the retrieved images shown in Figure 8 and two in Figure 9. For the rest of the retrieved images, ALFE demonstrates the ability to recognize the same or similar texture of the iris as well as a pupil similar in shape.

Table 4. Experimental results for ALFE efficiency evaluation

CASIA-IRIS-LAMP GROUPS	Relevant Retrieved Images	Irrelevant Retrieved Images	Retrieved Images	Precision (%)	Recall (%)
Group A	7	3	10	70	35
Group B	8	2	10	80	40
Group C	10	0	10	100	50
Group D	9	1	10	90	45
Group F	9	1	10	90	45
Group G	10	0	10	100	50
Group H	10	0	10	100	50
Group J	10	0	10	100	50
Group K	10	0	10	100	50
Group M	8	2	10	80	40
Group N	8	2	10	80	40
Group P	8	2	10	80	40
Group Q	10	0	10	100	50
Group S	8	2	10	80	40
Group Z	10	0	10	100	50

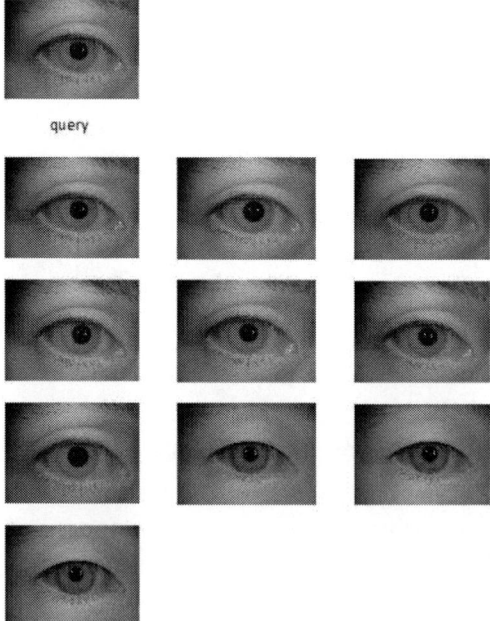

Figure 8. Extracted images upon request from Group A for ALFE.

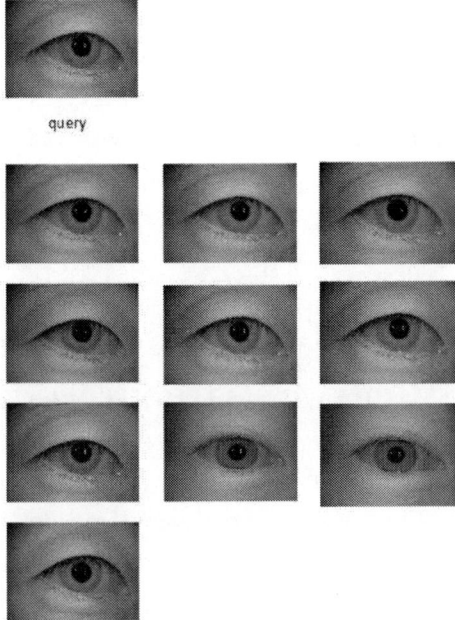

Figure 9. Extracted images upon request from Group M for ALFE.

The algorithm has the ability to recognize on the basis of alternating fluctuating brightness areas from lower to higher or vice versa intensity and with different size, the objects are centrally located in accordance with those in the query-image. ALFE has also the ability to recognize sharp jumps, the same complexity, repeatability and contrast, which influences the successful recognition of texture and shape.

When a query-image is submitted, ALFE demonstrates the ability to recognize radially spaced regions of the iris structure, similar to the query representing the iris structure, around the central object (pupil). The intensity of the respective pixels is characterized by both the same value according to the query and lower values. ALFE demonstrates the ability of 70% of the retrieved images (Figure 8) to recognize successfully the dark circle outlined around the iris according to the query-image. Successful recognition of transitions expressed in sharp jumps from higher to lower intensity and vice versa contributes to a high degree of recognition of the shape and texture of the iris, including details such as radial furrows, pupillary area, ciliary area, collarette and pupil size of the eye. When a radial furrows query-image is submitted, the algorithm under study demonstrates sensitivity to the distance between the upper and lower eyelid, retrieving images at the same or close distance from that of the query for 20% of the retrieved the results in Figure 8 and Figure 9. In addition, the ALFE shows slight sensitivity to the angle of the eye location relative to that of the query-image.

10. COMPARATIVE ANALYSIS ON EFFICIENCY EVALUATION WITHOUT RANK

The comparative analysis shows that the designed algorithm for content-based image retrieval by local features using Hausdorff distance (ALFH) enables content extraction by shape and texture of biometric data. The extraction accuracy when the Euclidean distance is used has lower values.

Analyzing the features of the four algorithms, it is found that ALPH is the best one for image recognition. The algorithm using local features and Euclidean distance (ALFE) and the algorithm using global features and Hausdorf distance (AGPH) demonstrate close opporunities with advantage of the former. The lowest result is demonstrated by the algorithm using global features and Euclidean Distance (AGFE).

The results obtained are due to the differences in the technological solution. Algorithms based on sub-image division in preprocessing and using local features provide the opportunity to retrieve more detailed and spatial information. This allows the recognition of images by texture and detail in the biometric data examined. Depending on the distance metric chosen, the result is better when applying Hausdorff distance than Euclidean distance.

The comparative analysis performed on the Precision and Recall metrics for the fifteen categories of the test image database shows the specificity of the four algorithms analyzed. The four algorithms demonstrate the best results in these metrics for the Group J and Group K categories. These results are explained by the presence of homogeneous regions with a smooth transition between high-intensity and low-intensity pixel zones and the

Group C, Group D, Group Q and Group Z have good values for the Precision criterion for the four algorithms. The results of Group A algorithms vary over a wide range, but it is observed that the ALFH result is the best one. For the other categories, all algorithms show intermediate results shown in Table 5.

The comparison between the algorithms by Precision proves that the designed algorithm using local features and Hausdorff distance (ALFH) demonstrates best results. The algorithms with global features (AGFH and AGFE) have the lowest values for this indicator.

In Figure 10 the correlation in the results of the Precision and Recall metrics for the fifteen categories of the four compared algorithms is illustrated.

Table 5. Experimental results for ALFH, ALFE, AGFH, AGFE efficiency evaluation

Image Category	ALFH		ALFE		AGFH		AGFE	
	Precision (%)	Recall (%)	Precision (%)	Recall (%)	Precision (%)	Recall (%)	Precision (%)	Recall (%)
Group A	80	40	70	35	50	25	50	25
Group B	70	35	80	40	70	35	70	35
Group C	100	50	100	50	70	35	100	50
Group D	100	50	90	45	100	50	100	50
Group F	70	35	90	45	90	45	90	45
Group G	100	50	100	50	100	50	90	45
Group H	100	50	100	50	100	50	80	40
Group J	100	50	100	50	100	50	100	50
Group K	100	50	100	50	100	50	100	50
Group M	100	50	80	40	70	35	70	35
Group N	80	40	80	40	100	50	80	40
Group P	100	50	80	40	70	35	80	40
Group Q	100	50	100	50	100	50	80	40
Group S	80	40	80	40	100	50	80	40
Group Z	90	45	100	50	100	50	100	50

On the other hand, the accomplished analysis regarding the invariance of brightness shows that all four algorithms are distinguished by this property when a query-image is submitted. The algorithms retrieve images with the group affiliation of the query-image with similar brightness distribution. Smooth transitions from darker to lighter, from left to right and upwards, as well as sharp jumps in the retrieved result according to the query, are observed. For Hausdorff distance algorithms, this property is more pronounced than those using Euclidean distance.

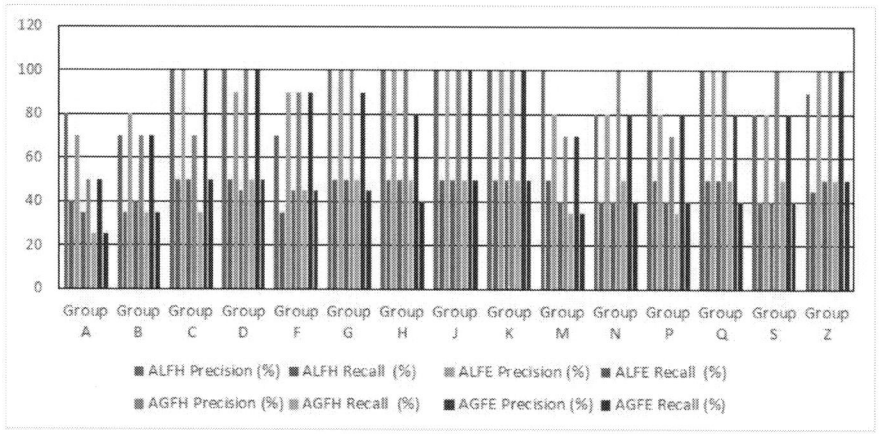

Figure 10. Ratio of Precision and Recall results for the fifteenth ALFH, ALFE, AGFH, AGFE categories.

AGFE stands out with the lowest values for time and image retrieval (t_{ImExt}), respectively at the highest speed. This result is due to the simplicity of the algorithm (with global features) as well as the simple mathematical formulation of the Euclidean distance used. In contrast, defining the similarity with Hausdorff's implementation requires a longer execution time.

Based on the total number of images extracted, I_{AVRRTR} is calculated. According to the result, AGFH shows the highest percentage of retrieved images. The smallest percentage of retrieved images is characterized by ALFE.

Table 6 lists the computed arithmetic mean values by the criteria for t_{ImExt} and I_{AVRRTR}.

Table 6. Arithmetic mean values by the criteria for t_{ImExt} and I_{AVRRTR} for ALFH, ALFE, AGFH, AGFE

Algorithm/Criteria	ALFH	ALFE	AGFH	AGFE
Image Retrieval Time (tImExt) (s)	0.49	0.41	0.37	0.21
Number of Retrieved Images (IAVR RTR)	383	219	583	367

Conclusion

Performance was tested using the classic method without rank using the Precision and Recall metrics. As a result of the studies and comparative analysis, ALFH demonstrates best results. It is found to be affected by the concentration of pixels of close or equal intensity in central areas (pupil and iris), which results in higher recognition accuracy by indicator "Group affiliation."

It has the ability to detect smooth transitions between areas of higher to lower intensity, as well as sharp transitions between areas of small size, equal complexity, repeatability and contrast, in texture detection. ALFH also has slight sensitivity to the orientation and arrangement of the elements.

The use of Hausdorff distance to compute similarity has a positive effect on shape recognition, and the application of DT CWT contributes to more accurate recognition of the iris texture.

Higher precision values are conditioned by the more detailed representation of the image by local feature vectors describing the spatial location of the elements in the biometric data.

References

[1] Jayarraman, K. B., S. Brigitha. J, "Survey on Content Based Image Retrieval Technique," *International Journal of Advanced Research in Computer Science and Software Engineering*, ISSN: 2277 128X, vol. 6, issue 3, pp. 587-591, 2016.

[2] Banuchitra, S., Dr. K. Kungumaraj, "A Comprehensive Survey of Content Based Image Retrieval Techniques," *International Journal Of Engineering And Computer Science*, ISSN: 2319-7242, vol. 5, issues 8, pp. 17577-17584. 2016.

[3] Barbu, T. "Content-based Image Retrieval using Gabor Filtering," *20th International Workshop Database and Expert Systems Application*, pp. 236-240, 2009.

[4] Shriwas, K., V. Ansari, "Content Based Image Retrieval using Model Approach," *International Journal of Applied Information Systems (IJAIS)*, vol. 10, no. 8, pp. 27-32, April 2016.

[5] Murugan, M. Vel., P. Jeyanthi, "Content Based Image Retrieval Using Color And Texture Feature Extraction In Android," *International Conference on Information Communication and Embedded Systems (ICICES)*, doi: 10.1109/ICICES.2014.7034033, pp. 1-7, 2014.

[6] Seung Gwan Lee, Daeho Lee, Youngtae Park, "Pupil Segmentation Using Orientation Fields, Radial Non-Maximal Suppression and Elliptic Approximation," *Advances in Electrical and Computer Engineering*, vol. 19, Number 2, 2019, pp. 69-74.

[7] Satish Rapaka, Y. B. N. V. Bhaskar, P. Rajesh Kumar, "An Efficient Method for Segmentation of Noisy and Non-circular Iris Images Using Optimal Multilevel Thresholding," *Journal of Advanced Research in Dynamical and Control Systems*, 11(2):1010-1022, July 2019.

[8] Chitra, P. L., Amsavalli Subramanian P., "Human Iris Pattern for Biometric Identification using Hough Transform with Daughman's Rubber Sheet model," *International Journal of Applied Engineering Research*, ISSN 0973-4562, vol. 10 no.76 (2015), pp. 388-393.

[9] Vimala, P., C. Karthika Pragadeeswari, "Secure Authentication with Iris using Hamming Distance," *International Journal of Recent Technology and Engineering (IJRTE)*, ISSN: 2277-3878, vol. 8, issue 4, November 2019, pp. 4787 – 4790.

[10] RachidaTobji, Wu Di, Naeem Ayoub, "A Synthetic Fusion Rule Based on FLDA and PCA for Iris Recognition Using 1D Log-Gabor Filter, " *Mathematical Problems in Engineering*, vol. 2019, article ID 7951320, 11 pages.

[11] Prajwala, N. B., N. B. Pushpa, "Matching of Iris Pattern Using Image Processing," *International Journal of Recent Technology and Engineering (IJRTE)* ISSN: 2277-3878, vol. 8, issue 2S11, September 2019, pp. 21-23.

[12] Rachida Tobji, Wu Di, Naeem Ayoub, Samia Haouassi, "Efficient Iris Pattern Recognition Method by using Adaptive Hamming Distance and 1D Log-Gabor Filter, " *(IJACSA) International Journal of Advanced Computer Science and Applications*, vol. 9, no. 11, 2018, pp. 662-669.

[13] Abhishek Gangwar, Akanksha Joshi, Padmaja Joshi, R. Raghavendra, "*DeepIrisNet2: Learning Deep-IrisCodes from Scratch for Segmentation-Robust Visible Wavelength and Near Infrared Iris Recognition*", 2019, pp.1-10.

[14] Wei Zhang, Xiaoqi Lu, Yu Gul, Yang Liu, Xianjing Meng, Jing Li, "*A Robust Iris Segmentation Scheme Based on Improved U-Net,* " doi: 10.1109/ACCESS.2019.2924464, IEEE Access, 2019.

[15] Yaosong Cheng, Yuanning Liu, Xiaodong Zhu, And Shuo Li, "*A Multiclassification Method for Iris Data based on the Hadamard Error Correction Output Code and a Convolutional Network,* " doi: 10.1109/ACCESS.2019.2946198, IEEE Access, 2017 https://ieeexplore.ieee.org/stamp/stamp.jsp?arnumber=8862846.

[16] Nour Eldeen M. Khalifa, Mohamed Hamed N. Taha, Aboul Ella Hassanien, Hamed Nasr Eldin T. Mohamed, "Deep Iris: Deep Learning for Gender Classification Through Iris Patterns, " *Acta Informatica Medica*, vol. 27, issue 2, 2019 Jun 27(2), pp. 96-102.

[17] Hylish James, VarshaVishwakarma, "Iris Recognition Using CNN With Normalization Technique," *IJRAR19K6911 International Journal of Research and Analytical Reviews (IJRAR)*, 2019 IJRAR June 2019, vol. 6, issue 2, pp.433-436.

[18] Shideh Homayon, Mahdi Salarian, "*Iris Recognition For Personal Identification Using Lamstar Neural Network*," arXiv:1907.12145 [cs.CV], 2019.

[19] Vimala, P., C. Karthika Pragadeeswari, "Secure Authentication with Iris using Hamming Distance," *International Journal of Computer*

Science and Information Security (IJCSIS), vol. 15, no. 9, September 2017, pp. 278-280.

[20] Kingsbury N. G.: "Complex wavelets for shift invariant analysis and filtering of signals," *Journal of Applied and Computational Harmonic Analysis*, vol 10, no 3, pp. 234-253, May 2001.

[21] Kingsbury, N., A. Zymnis, "3D DT-MRI Data Visualisation using the Dual Tree Complex Wavelet Transform," in *Proceedings EURASIP Biosignal Conference*, Brno Czech Rep., June 2004, paper 31.

[22] Kingsbury, N., A. Zymnis, Alonso Pena, "DT-MRI Data Visualisation Using The Dual Tree Complex Wavelet Transform," *3rd IEEE International Symposium on Biomedical Imaging: Macro to Nano*, vol. 111, pp. 328331, Los Alamitos 2004.

[23] Chen, H., N Kingsbury, "Efficient Registration of Nonrigid 3-D Bodies," in *IEEE Transactions On Image Processing*, vol. 21, no. 1, pp. 262-272, January 2012.

[24] Hong, T., N. Kingsbury, "Estimation of the Fundamental Matrix Based on Complex Wavelets," *International Conference on Networking and Information Technology* (ICNIT), DOI: 10.1109/ICNIT.2010. 5508498, pp. 350-354, July 2010.

[25] Kingsbury, N. "A Dual-Tree Complex Wavelet Transform with Improved Orthogonality And Symmetry Properties," in *Proceedings International Conference on Image Processing*, pp. 375-378, 2002.

[26] Cui, D., Y. Liu, J. Zuo, Bo Xu, "A Modified Image Retrieval Algorithm Based on DTCWT," in *Journal of Computational Information Systems*, vol.7, no.3, pp. 896-903, 2011.

[27] Chauhan, R., R. Dwivedi, R. Bhagat, "Comparative Analysis of Discrete Wavelet Transform and Complex Wavelet Transform For Image Fusion and De-Noising," in *International Journal of Engineering Science Invention,* vol. 2, issue 3, March 2013 pp. 18-27.

[28] Selesnick, I., R. Baraniuk, N. Kingsbury, "The Dual-Tree Complex Wavelet Transform," in *IEEE Signal Processing Magazine*, pp. 123-151, November 2005.

[29] *CASIA Iris Image Database*, http://biometrics.idealtest.org/

[30] Vetova, S., I. Ivanov, "Efficiency Comparative Analysis Between Two Search Algorithms Using DT CWT For Content-Based Image Retrieval," in *Proceedings of the 2015 IEEE International Conference on Industrial Technology (ICIT 2015),* Seville, Spain, March, 2015, pp. 1664-1669, IEEE Catalog Number: CFP15CIT-USB, ISBN: 978-1-4799-7799-4, 978-1-4799-7799-4/15 ©2015 IEEE.

[31] Vetova, S., I., Ivanov, "Image Features Extraction Using The Dual-Tree Complex Wavelet Transform", *2nd International Conference on Mathematical, Computational And Statistical Sciences*, Gdansk, Poland, 2014, pp. 277 - 282.

In: The Fundamentals of Search Algorithms
Editor: Robert A. Bohm

ISBN: 978-1-53619-007-6
© 2021 Nova Science Publishers, Inc.

Chapter 3

DIFFERENTIAL EVOLUTION FOR SOLVING CONTINUOUS SEARCH SPACE PROBLEMS

Omar Andres Carmona Cortes[*] *and Hélder Pereira Borges*[†]
Computer Department (DComp)
Instituto Federal de Educao,
Ciłncia e Tecnologia do Maranho (IFMA),
Monte Castelo, So Luis, MA, Brazil

Abstract

Continuous search space problems are difficult problems to solve because the number of solutions is infinite. Moreover, the search space gets more complex as we add constraints to the problem. In this context, this chapter aims to show the usage of the differential evolution algorithm for solving continuous search space problems using unconstrained functions and a constrained real-world problem. Six different mutation strategies were implemented: /DE/Rand/01, /DE/Best/01, /DE/Rand/02, /DE/Best/02, /DE/Rand-to-Best/01, and /DE/Rand-to-Best/02. These strategies were tested in five unconstrained continuous benchmark functions with different features and complexity of search space: Rosenbrock, Sphere, Schwefel, Rastrigin, and Griewank. Also, a problem called the economic dispatch problem, whose system comprises 40 generators, was optimized. To compare the strategies, we used a Kruskal-Wallis H-Test. Then we used the pairwise Wilcox test to discover where the differences

[*]Corresponding Author's Email: omar@ifma.edu.br.
[†]Corresponding Author's Email: helder@ifma.edu.br.

are. Results have shown that strategies /DE/Best/01 and /DE/Rand-to-Best/01 tend to present the best outcomes. While in the economic dispatch problem, the winner strategy varies as we increase the number of iterations, probably because eventually all strategies can reach a good solution.

1. INTRODUCTION

In the optimization context, the search space and its properties are defined based on how knowledge is represented. In this way, describing challenging problems search space can help in building good solutions. Usually, search spaces could be broken up into states, and the problem space description should describe how to move from one state to the next. Thus, discrete spaces have a discrete number of states, eventually infinite, while continuous spaces have uncountable states.

Considering how the search space looks, particular search algorithms could be used to find the answer by moving through the search space. Further, it can be used to infer, discuss, or prove aspects of the problem. Furthermore, selecting a suitable search space may be more relevant than the definition of a solving algorithm. Therefore, there may be an advantage in using a continuous search space over a discrete one that could be verified based on the quality of the final solution and stability of intermediate partial solutions.

Anyway, some algorithms for optimization problems were initially conceived and developed, considering only discrete optimization problems. Thus, for a correct application in continuous optimization problems, it is necessary to transform the continuous search space into a discrete one, discretizing the continuous decision variables, resulting in a discrete set of values in which the search for an optimal solution can be conducted [1]. However, continuous spaces can produce sequential steps that do not improve the result, converging to undesired limit points; thus, due to the discretization of the search space, the algorithms' performance in continuous problems may not be considered suitable.

Even though an approach performing a discrete refinement can result in a good solution in the continuous search space for small problems, this is not true in real-world problems because, given particular characteristics and complexities, these can usually present a vast space for continuous search. Also, it is quite common to lose the global optima domain in the early stages of refinement. The generation of a non-homogeneous and somewhat random mesh throughout the search space can minimize this problem. Further, it can be efficient to deal with the combination of discrete and continuous decision variables.

In this context, evolutionary and swarm computation appears as attractive approaches to tackle this kind of problem/space. One of these algorithms is the Differential Evolution (DE) [2] that works similarly to Genetic Algorithms (GA) [3], but using a different order in the genetic operators. While in GA the operators are crossover and mutation, in DE the order is reverse performing mutation and then crossover.

DE has already been applied to solve problems in different scientific areas; thus, it is possible to verify in the literature applications in medicine, industry, economics, biotechnology, acoustics, transport, environmental sciences, and aerodynamics, electrical engineering, and others [4]. This diversity of applications using this algorithm represents an excellent indication of its versatility and efficiency. Moreover, DE is an evolutionary approach capable of dealing with non-linear, non-differentiable and multimodal objective functions. Usually, its performance overcomes several optimization algorithms in terms of robustness and speed of convergence take into account benchmark of common problems and real-world applications [5], [2].

Despite the robustness and reliability performance of DE, another possible reason behind his popularity is the simplicity of its structure, facilitating the understanding of its technique. The canonical DE needs to configure only three parameters: amplification factor (F), crossover rate (CR), and population size P. Additionally, we can also choose which kind of strategy the algorithm will use, being the most common /DE/Rand/01 and /DE/Best/01.

Hence, in this work, we evaluate how ED behaves on solving continuous optimization problems using six different strategies: /DE/Rand/1, /DE/Best/1, /DE/Rand/2, /DE/Best/2, /DE/rand-to-best/1, and /DE/rand-to-best/2. Therefore, the idea is to evaluate five benchmark functions with different features and a real-world problems called economic dispatch. In this last application, the algorithm will be tested in an actual power plant configuration composed of forty generators. For this sake, this work is divided as follows: Section 2 introduces the central concept of continuous search spaces and formulates all considered problem; Section 3 shows the Differential Evolution Algorithms and its strategies; Section 4 presents the experiments and the statistical evaluations; finally, Section 4.4 show the conclusions of this work.

2. CONTINUOUS SEARCH SPACE

A continuous search space belongs to the $\mathbb{R}$ space, meaning that every parameter or variable also belong to $\mathbb{R}$. In other words, each variable can assume an infinity possibilities within a certain domain. Mathematically, we have $f(x)$ such that x is a vector in which $a \leq x_i \leq b \mid x_i \in \mathbb{R}$, *i.e.*, any variable must be in the interval $[a, b]$.

Three other properties are crucial in numerical optimization: dimensionality, separability and modality. Dimensionality concerns the number of parameters or variables. The more variables the problem has, the more difficult is to solve it. In other words, as the number of variables increases, the search space's complexity increases exponentially [6].

The separability involves the possibility of dividing $f(x)$ into two or more functions. Thus, If all the parameters or variables are independent, then a sequence of n independent optimization processes can be performed [7]. Consequently, inseparable functions are more challenging to optimize then separable ones because variables can depend each other in inseparable functions. Modality concerns the existence or not of many local optima. Thus, inseparable and multimodal functions represent a more significant challenge to solve than the other ones.

As previously mentioned, we will test our code using the following five unconstrained continuous numerical benchmarks functions: Rosenbrock, Sphere, Schwefel, Rastrigin, and Griewank. Table 1 presents the function formulations and Table 2 shows benchmarks properties (Domain, Global Optima (Min), Separability and Modality). Even though these benchmark functions are considered unconstrained, they are abiding by domain constraint, i.e., each variable is narrowed by lower and upper boundaries.

Table 1. Unconstrained Benchmark Functions

Name	Function
Rosenbrock	$f_1(x) = \sum_{i=1}^{n} [100(x_{i+1} - x_i^2)^2 + (x_i - 1)^2]$
Sphere	$f_2(x) = \sum_{i=1}^{n} x_i^2$
Schwefel	$f_3(x) = \sum_{i=1}^{n} -x_i \sin \sqrt{|x_i|}$
Rastrigin	$f_4(x) = 10n + \sum_{i=1}^{n} (x_i^2 - 10 \cos(2\pi x_i))$
Griewank	$f_5(x) = \frac{1}{4000} \sum_{i=1}^{n} x_i^2 - \prod_{i=1}^{n} \cos(\frac{x_i}{\sqrt{i}})$

Table 2. Benchmark functions properties

Name	Domain	Min	Separable	Multimodal
Rosenbrock	[-2.048, 2.048]	0	No	No
Sphere	[-5.12, 5.12]	0	Yes	No
Schwefel	[-500, 500]	-12569.5	Yes	Yes
Rastrigin	[-5.12, 5.12]	0	Yes	Yes
Griewank	[-600, 600]	0	No	Yes

The first two functions are unimodal and usually are chosen to be a first step in evaluating optimization problems. The first one, Rosenbrock, is also an inseparable function and represents a challenging function due to its quadratic characteristic. Some authors also consider it as multimodal in high dimensions, as presented in al-Rifaie's work [8]. Either, Rosenrock is considered a challenge in the numerical optimization field [9] [10]. The second function is the Sphere function, a unimodal and symmetric one, which is also used to derive other functions such as the Rastrigin one.

The next three functions are multimodal. Schwefels function, also known as Surface Schwefel, is composed of a great number of peaks and valleys. The function has a second-best minimum far from the global minimum, where many search algorithms are trapped. Furthermore, the global minimum is near the bounds of the domain [11]. As previously mentioned, the Rastrigins function was constructed from Sphere. Its contour is made up of many local minima, whose value increases with the distance to the global minimum. Therefore, the algorithm could be trapped if genes are produced far from the global optima but into the domain.

Finally, Griewanks function has a product term that introduces interdependence among the variables. The aim is the failure of the techniques that optimize each variable independently. Moreover, the optima of Griewanks function are regularly distributed. Next, we present the formalization of a real-world problem called the economic dispatch problem, which also constitutes a continuous search space problem.

2.1. The Economic Dispatch Problem

A power system is a network consisting of generation, distribution, and transmission systems. It is a complex system subject to a wide variety of physical and operational restrictions, making their planning, control, and operation a great challenge, mainly due to the need to determine an ideal state, which satisfies the many constraints relevant in this area. In this segment, we have the Economic Dispatch Problem (EDP), one of the vital optimization problems in power system operation, playing a critical role in the operational process and planning of these systems, having received significant attention in recent years.

The primary purpose of the EDP [12] [13] is to minimize the total power production costs while generation constraints are satisfied. In the traditional solution, an approximate quadratic function is used to make the mathematical formulation of the EDP convex so that it is possible to reduce the computational overhead. However, in practice, on the one hand, the machines and tools used can make the input-output curves of the generators have highly non-linear characteristics.

The non-linearity and discontinuities of power systems usually come from the effect of "valve points" [14], making optimization more challenging to treat. In this context, conventional solution methods, such as Lagrange relaxation, non-linear programming, linear programming, dynamic programming, quadratic programming, and interior point method will have greater difficulties, because in general, they present a strong imposition of several restrictions, such as convexity, continuity, and differentiability in the objective functions. Also, they are highly sensitive to the initial values of the optimized variables involved in the procedure.

As a viable alternative to these methods, conventional methods, there are also metaheuristic methods, which have attracted considerable attention in recent years. They do not have strict requirements regarding the formulation of optimization problems and can avoid the influences of the initial condition's sensitivity and gradient information.

Thus, EDP belongs to a class of optimization methods of non-linear programming that contains equality and inequality restrictions and must provide an ideal generation dispatch between the operational units, satisfying the system load demands and the practical operating limits for generators. In other words, cost efficiency is the most crucial subproblem of power system operations. Thus, the generation cost function is usually expressed as a quadratic polynomial as expressed by Equation 1, in which F is the cost, P_i is the power produced by generator i, and a, b, and c are generation constants.

$$F(P_i) = \sum_{i=1}^{n} a_i + b_i P_i + c_i P_i^2 \qquad (1)$$

Additionally to the cost of generations, we have the valve point effect, which is modeled as $e_i \sin f_i (P_{min} - P_i)$, adding a non-linearity to the problem. Consequently, the cost of generating power can be rewritten as expressed in Equation 2, in which e_i and f_i are constants of the valve effect, and $P_{i\,min}$ is the lower bound of energy production for genertor i.

$$F(P_i) = \sum_{i=1}^{n} a_i + b_i P_i + c_i P_i^2 + |e_i \sin f_i \times (P_{i\,min} - P_i)| \qquad (2)$$

Therefore, the optimization problem in the continuous search space can be written as presented in Equation 3, whose purpose is to minimize the cost of producing power considering the valve point effect.

$$Min\ F(P_i) = Min \sum_{i=1}^{n} a_i + b_i P_i + c_i P_i^2 + |e_i \sin f_i (P_{i\,min} - P_i)| \qquad (3)$$

Having said that, we can express the main constraint of the problem in Equation 4, in which P_D is the demanded power. In other words, the produced power must be greater or equal to the required energy, also known as load demand.

$$F(P_i) \geq P_D \qquad (4)$$

Regardless the load demand, generators can not operate in any range. They are subjected to generation constrains as presented in Equation 5. As we can notice, each generator P_i has its operation limits between P_{min} and P_{max}, making the search space a much more complex one.

$$P_{min} \leq P_i \leq P_{max} \qquad (5)$$

Next, we detail how the differential evolution works and how constraints can be tackled.

3. DIFFERENTIAL EVOLUTION

Storn and Price [5] presented the Differential Evolution (DE) in 1997. It is a simple and efficient scheme for global optimization over continuous spaces, being considered a stochastic optimizer of functions based on the population of possible solutions. It is similar to the Genetic Algorithm (GA); however, mutation comes first and then crossover in DE. Despite being classified as an evolutionary algorithm, DE has no basis or inspiration in any natural process.

The basic approach of ED is to generate a completely random population. In this context, a population is a set of solutions to a specific problem. From this population, vectors with experimental parameters of differences must be generated, created from pairs of vectors of the population itself, and then used to carry out a mutation in the population vector. This perturbation vector is created from the random selection of three individuals from the current population, two of which will be used to generate a disturbance parameter calculated by the difference between them. This result is then added to the third individual. In this way, this new mutant solution is formed, the result of a disturbance in one of the individuals of the original population.

The next step is to perform a crossover between individuals from the population and the mutant population, generating new offspring. Lastly, the selection process will be carried out; thus, if the descendant's fitness is better than the original, it should remain for the next generation taking the place of the target in the population.

The Algorithm 1 shows its pseudo-code, in which a random population is created then evaluated. The algorithm continues while the stop criterion is not reached. In this loop, the algorithm starts the mutation operation by randomly selecting three individuals and creates a vector of differences v. Afterward, the crossover is performed by creating a new individual selecting a gene from v or the current individual. If the new individual is better than the current one, it replaces it.

As all individuals are randomly selected when creating v in the DE's pseudo-code, the mutation strategy is called /DE/Rand/1. However, other approaches to generate the mutation vector can be used. For example, if $indiv_1$ is replaced by the best solution in the population, the name of strategy changes to /DE/Best/1. Hence, the strategy's name tends to follow how individuals are selected to create the mutation vector. In this context, we can cite some other strategies as follows:

Algorithm 1 - Differential Evolution Pseudo-Code

pop ← InitPopulation();
fitness ← Eval(pop);
$i \leftarrow 1$
while stop Criterion not reached **do**
    Select 3 individuals randomly: $indiv_1, indiv_2, indiv_3$;
    $v_j \leftarrow indiv_3 + F \times (indiv_1 - indiv_2)$;
    if (rand() ¡ CR) **then**
        $new\_indiv_j \leftarrow v_j$
    else
        $new\_indiv_j \leftarrow pop_{ij}$
    end if
    if fitness(new_indiv) better than fitness(pop_i) **then**
        $pop_i \leftarrow new\_indiv$;
    end if
    $i++$
end while

- /DE/Rand/01: $v_j \leftarrow indiv_1 + F \times (indiv_2 - indiv_3)$

- /DE/Rand/02: $v_j \leftarrow indiv_1 + F \times (indiv_2 - indiv_3) + F \times (indiv_2 - indiv_3)$

- /DE/Best/01: $v_j \leftarrow best + F \times (indiv_1 - indiv_2) + F \times (indiv_2 - indiv_3)$

- /DE/Best/02: $v_j \leftarrow best + F \times (indiv_1 - indiv_2) +$

- /DE/Rand-to-Best/01: $v_j \leftarrow indiv_1 + F \times (best - indiv_2)$

- /DE/Rand-to-Best/02: $v_j \leftarrow indiv_1 + F \times (best - indiv_2) + F \times (indiv_3 - indiv_4)$

As we can see, the primordial differences between the referred strategies are how the individuals are chosen and how many differences are computed. For instance, in the method /DE/Rand-to-Best/02, the main individual ($indiv_1$) is selected randomly, two vectors of differences are taking place, and the best solution is used only in the first vector. Next, we explain how we will deal with constraints.

3.1. Handling Constraints

Constraints are rules that the problem is abided by to produce valid solutions. Therefore, it is essential to define approaches to tackle them. The most radical way of dealing with constraint is the death penalty. In this approach, a violation in a constraint leads to excluding the individual from the population. This approach's drawback is that the probability of generating a low number of individuals is high if the problem has many constraints.

Handling the power generation constraint (Equation 5) is relatively simple because we have to keep each variable within a domain. Thus, after creating each v_i we control the limits of generation. If a limit is violated, we saturate the value on maximum or minimum, depending on the violation's direction. If a generator produces less power than the lower bound, we replace the value with the lower bound, and if the value is higher than the upper bound, we assign the upper bound to the variable. Alternatively, we can create a generation value within the domain randomly.

In this context, a common way of tackling constraint is the static penalty. Thus, a penalty is added to the objective function for any violations in the constraints. Mathematically, the static penalty is shown in Equation 6, in which x is a solution, $p(x)$ is a penalty function, and F is the feasible search space.

$$fitness = \begin{cases} f(x), & \text{If } x \in F \\ f(x) + p(x), & \text{If } x \notin F \end{cases} \quad (6)$$

4. EXPERIMENTS

4.1. Setup

The experiments were executed in Intel i5 9th generation with 2.4Ghz, 8GB of RAM. The programming language was R version 3.6.0 64 bits under Windows 10 also 64 bits. All functions were executed 30 times using 500, 1000, and 2000 iterations. The Differential Evolution was composed of a population equal 50, dimension equals 30 in the benchmark functions and 40 in the EDP, F equals 0.6, and CR equals 0.8. Because we did not test the experiments' distribution, the comparison was performed using a Kruskall-Wallis [15] H-test. The null hypothesis (H_0) is that there are no meaningful differences between the strategies. Therefore, the alternative hypothesis (H_1) is that the strategies are different. A pairwise Wilcox test [15] with a significance of 95% was performed to discover where the differences between strategies are. In the Wilcox test we used the Bonferroni correction that is the standard p-value correction.

The benchmark functions were the same presented in Table 1 following the domain shown in Table 2. Concerning the economic dispatch problem, we used a system devised by 40 generators whose constants and operation boundaries are presented in Table 3. Regardless of the problem, results always present the minimum (min), the maximum (max), the mean, and the standard deviation (SD). The strategies follows the following legend in tables: 1 - /DE/Rand/01; 2 - /DE/Best/01; 3 - /DE/Rand/02; 4 - /DE/Best/02; 5 - /DE/Rand-to-Best/01; and 6 - /DE/Rand-to-Best/02.

Next, we explain how DE treated the domain constraint and the economic dispatch constraints.

4.2. Handling Constraints

In benchmark functions, the only constraint is the domain one. Thus, we keep each dimension's values by saturating the minimum or maximum when the boundaries are violated. Creating a number within the domain when it is violated is also possible; however, saturating the values led to better results.

In the economic dispatch problem, we have one more constraint to attend to a required demand. We tackled it using a penalty function as presented in Equation 6, in which $p(x) = 2 \times [(\sum_i^G P_i) - P_D)]^3$ where G is the number of generators and P_D is the power demand. In other words, the penalty is twice times the cube of the difference between the produced power and the demand. Consequently, we can transform Equation 6 into Equation 7 as follows, in which F is the feasible search space:

$$fitness = \begin{cases} f(x), & \text{If } x \in F \\ f(x) + 2 \times (\sum_i^G P_i - P_D)^3, & \text{If } x \notin F \end{cases} \quad (7)$$

In the next section, we present the benchmarks function results, in which only the domain constraint is considered.

4.3. Results for Benchmark Functions

Table 4 shows the results using 500 iterations, in which we can see that the strategy /DE/Best/01 presents the best results (min) for Rosenbrock, Sphere,

Table 3. Power system composed of 40 generators

#Gen	a	b	c	e	f	P_{min}	P_{max}
1	0.01	9.73	94.70	100.00	0.08	36.00	114.00
2	0.01	9.73	94.70	100.00	0.08	36.00	114.00
3	0.02	7.07	309.54	100.00	0.08	60.00	120.00
4	0.01	8.18	369.03	150.00	0.06	80.00	190.00
5	0.01	5.35	148.89	120.00	0.08	47.00	97.00
6	0.01	8.05	222.33	100.00	0.08	68.00	140.00
7	0.00	8.03	287.71	200.00	0.04	110.00	300.00
8	0.00	6.99	391.98	200.00	0.04	135.00	300.00
9	0.01	6.60	455.76	200.00	0.04	135.00	300.00
10	0.01	12.90	722.82	200.00	0.04	130.00	300.00
11	0.01	12.90	635.20	200.00	0.04	94.00	375.00
12	0.01	12.80	654.69	200.00	0.04	94.00	375.00
13	0.00	12.50	913.40	300.00	0.04	125.00	500.00
14	0.01	8.84	1760.40	300.00	0.04	125.00	500.00
15	0.01	9.15	1728.30	300.00	0.04	125.00	500.00
16	0.01	9.15	1728.30	300.00	0.04	125.00	500.00
17	0.00	7.97	647.85	300.00	0.04	220.00	500.00
18	0.00	7.95	649.69	300.00	0.04	220.00	500.00
19	0.00	7.97	647.83	300.00	0.04	242.00	550.00
20	0.00	7.97	647.81	300.00	0.04	242.00	550.00
21	0.00	6.63	785.96	300.00	0.04	254.00	550.00
22	0.00	6.63	785.96	300.00	0.04	254.00	550.00
23	0.00	6.66	794.53	300.00	0.04	254.00	550.00
24	0.00	6.66	794.53	300.00	0.04	254.00	550.00
25	0.00	7.10	801.32	300.00	0.04	254.00	550.00
26	0.00	7.10	801.32	300.00	0.04	254.00	550.00
27	0.52	3.33	1055.10	120.00	0.08	10.00	150.00
28	0.52	3.33	1055.10	120.00	0.08	10.00	150.00
29	0.52	3.33	1055.10	120.00	0.08	10.00	150.00
30	0.01	5.35	148.89	120.00	0.08	47.00	97.00
31	0.00	6.43	222.92	150.00	0.06	60.00	190.00
32	0.00	6.43	222.92	150.00	0.06	60.00	190.00
33	0.00	6.43	222.92	150.00	0.06	60.00	190.00
34	0.00	8.95	107.87	200.00	0.04	90.00	200.00
35	0.00	8.62	116.58	200.00	0.04	90.00	200.00
36	0.00	8.62	116.58	200.00	0.04	90.00	200.00
37	0.02	5.88	307.45	80.00	0.10	25.00	110.00
38	0.02	5.88	307.45	80.00	0.10	25.00	110.00
39	0.02	5.88	307.45	80.00	0.10	25.00	110.00
40	0.00	7.97	647.83	300.00	0.04	242.00	550.00

Rastrigin, and Griewank. While the strategy /DE/Rand-to-Best/01 shows the best outcome in Rastrigin's function.

Table 4. Benchmarks Functions with dimension equals 30 and 500 iterations

	Rosenbrock					
	1	2	3	4	5	6
Min	24.64	**0.57**	204.71	26.61	22.05	67.73
Max	26.77	79.81	417.09	37.71	80.03	191.15
Mean	26.32	23.04	304.28	30.63	25.84	122.08
SD	0.37	11.58	52.08	2.87	10.26	31.01
	Sphere					
	1	2	3	4	5	6
Min	1.9004e-05	**3.4710e-20**	1.2260e+00	1.4817e-03	3.3327e-09	3.0298e-01
Max	6.8450e-05	3.4772e-18	4.7655e+00	1.2554e-02	2.7091e-08	1.1073e+00
Mean	3.8950e-05	7.4563e-19	2.8984e+00	5.5242e-03	1.0162e-08	5.6509e-01
SD	1.2260e-05	8.3486e-19	6.9796e-01	2.6955e-03	5.6458e-09	1.8802e-01
	Schwefel					
	1	2	3	4	5	6
Min	-9.2471e+03	-1.0688e+04	-6.4664e+03	-9.8386e+03	**-1.1785e+04**	-6.7976e+03
Max	-7.6052e+03	-8.0536e+03	-5.5004e+03	-5.9863e+03	-8.5067e+03	-5.4447e+03
Mean	-8.4758e+03	-9.7498e+03	-5.9014e+03	-7.9675e+03	-9.8980e+03	-6.0167e+03
SD	3.1589e+02	5.8361e+02	2.1979e+02	9.5229e+02	7.2976e+02	3.2126e+02
	Rastrigin					
	1	2	3	4	5	6
Min	75.23	**31.84**	189.64	147.95	45.73	202.78
Max	114.82	79.60	250.41	211.74	111.82	253.82
Mean	95.56	56.27	225.91	183.03	84.55	226.96
SD	10.29	12.38	15.94	16.70	12.75	12.54
	Griewank					
	1	2	3	4	5	6
Min	1.4789e-02	**0.0000e+00**	7.5690e+00	8.0461e-01	2.2776e-06	1.8086e+00
Max	5.2754e-01	3.4477e-02	1.5929e+01	1.0535e+00	1.2326e-02	4.0507e+00
Mean	8.3465e-02	7.5554e-03	1.0973e+01	9.9121e-01	1.6832e-03	2.7682e+00
SD	9.9975e-02	9.3956e-03	2.3924e+00	5.9990e-02	3.4325e-03	5.3574e-01

The Kruskal-Wallis test presented a p-value equal to $2.2e^{-16}$ in all benchmark functions, indicating strong evidence that the difference between algorithm using 500 iterations exists. Thus, Table 5 presents a pairwise Wilcox test to indicate where the differences are. As we can see, because p-values from the table are less than 0.05, there are differences mostly between all algorithms excepting the pairs /DE/Rand-to-Best/01 - /DE/Best/01 and /DE/Rand-to-Best/02 - /DE/Rand/02 in Schwefel. Either, there are no differences between the pairs /DE/Rand-to-Best/02 - /DE/Rand/02 and /DE/Best/02 - /DE/Rand/01 in Griewank. As the best values do not belong to these pairs, the results of the best strategies still those indicated in Table 4.

Table 5. Pairwise Wilcox Test - 500 it

	Rand-to-Best/01	Rand-to-Best/02	Best/01	Best/02	Rand/01
Rosenbrock					
Rand-to-Best/02	1.0147e-15	-	-	-	-
Best/01	6.3226e-06	1.0147e-15	-	-	-
Best/02	7.2623e-12	2.5367e-16	2.0574e-11	-	-
Rand/01	2.5124e-11	2.5367e-16	1.1251e-07	4.8197e-15	-
Rand/02	2.5367e-16	2.5367e-16	2.5367e-16	2.5367e-16	2.5367e-16
Sphere					
Rand-to-Best/02	2.5367e-16	-	-	-	-
Best/01	2.5367e-16	2.5367e-16	-	-	-
Best/02	2.5367e-16	2.5367e-16	2.5367e-16	-	-
Rand/01	2.5367e-16	2.5367e-16	2.5367e-16	1.0865e-08	-
Rand/02	2.5367e-16	2.5367e-16	2.5367e-16	2.5367e-16	2.5367e-16
Schwefel					
Rand-to-Best/02	1.9386e-10	-	-	-	-
Best/01	**0.1647**	2.3581e-12	-	-	-
Best/02	3.5260e-14	2.5367e-16	2.5367e-16	-	-
Rand/01	2.5367e-16	2.5367e-16	2.5367e-16	2.5367e-16	-
Rand/02	2.3060e-10	**1.00**	3.7174e-11	2.5367e-16	2.5367e-16
Rastrigin					
Rand-to-Best/02	2.5367e-16	-	-	-	-
Best/01	2.5367e-16	9.1733e-03	-	-	-
Best/02	2.9760e-12	2.5367e-16	2.5367e-16	-	-
Rand/01	2.5367e-16	2.5367e-16	2.5367e-16	2.5367e-16	-
Rand/02	2.5367e-16	1.8987e-09	1.7757e-15	2.5367e-16	2.5367e-16
Griewank					
Rand-to-Best/02	2.5367e-16	-	-	-	-
Best/01	2.5367e-16	2.5367e-16	-	-	-
Best/02	2.5367e-16	2.5367e-16	2.5367e-16	-	-
Rand/01	8.8936e-13	2.5367e-16	2.5367e-16	**0.7183**	-
Rand/02	4.4705e-10	**1.00**	1.3322e-09	4.4705e-10	4.4705e-10

Table 6 present the results for 1000 iterations. As we can see, the results are similar to the previous one except that /DE/Rand-to-Best/01 also reached the best result in Griewank. Thus, let us look at the Wilcox test in Table 7 to verify the differences in which we reject H_0. In Rosenbrock, only the pair

/DE/Rand/01 and DE/Best/02 present similar results. In the Sphere function, the performance of all strategies is different. In Schwefel, only the pair /DE/Rand-to-Best/01 - /DE/Rand/02 shows similar outcomes. In Rastrigin, the similarities are between DE/Best/01 and /DE/Rand/02. Finally, in Griewank, similarities are between the pairs /DE/Rand-to-Best/01 - /DE/Rand/02, /DE/Rand-to-Best/02 - /DE/Rand/02, and /DE/Best/01 - /DE/Rand/02. Some of this comparison can be considered odd, and they are probably due to Bonferroni correction. Thus, we still consider the best results as coming from /DE/Best/01 and /DE/Rand-to-Best/01.

Table 6. Benchmarks Functions with dimension equals 30 and 1000 iterations

	Rosenbrock					
	1	2	3	4	5	6
Min	22.86	**9.06**	36.64	22.70	16.08	26.35
Max	27.49	18.71	81.63	25.47	22.60	30.85
Mean	23.57	13.46	56.31	23.52	18.60	28.33
SD	0.79	2.29	12.44	0.65	1.26	0.86
	Sphere					
	1	2	3	4	5	6
Min	5.7732e-12	**5.9518e-41**	3.4479e-02	1.4196e-08	7.7880e-20	8.3745e-04
Max	4.6227e-11	3.1659e-36	9.6042e-02	6.4203e-07	3.2508e-18	4.2617e-03
Mean	1.6980e-11	1.0868e-37	5.8506e-02	2.3253e-07	7.2623e-19	2.2309e-03
SD	9.4581e-12	5.7742e-37	1.6745e-02	1.4735e-07	7.8944e-19	8.0562e-04
	Schwefel					
	1	2	3	4	5	6
Min	-1.1946e+04	-1.0589e+04	-7.0956e+03	-1.0904e+04	**-1.2093e+04**	-7.0920e+03
Max	-9.9476e+03	-8.4105e+03	-5.9259e+03	-8.8751e+03	-1.0588e+04	-6.0977e+03
Mean	-1.0777e+04	-9.6840e+03	-6.4271e+03	-9.5710e+03	-1.1369e+04	-6.5201e+03
SD	5.0165e+02	6.1492e+02	2.8473e+02	4.9465e+02	3.9181e+02	2.3075e+02
	Rastrigin					
	1	2	3	4	5	6
Min	33.63	25.87	176.32	108.31	**25.16**	158.98
Max	54.35	73.65	212.23	183.63	56.97	216.13
Mean	45.43	50.26	195.60	139.13	38.73	187.95
SD	5.11	13.94	9.69	19.49	7.58	14.12
	Griewank					
	1	2	3	4	5	6
Min	4.6668e-09	**0.0000e+00**	1.1173e+00	5.0528e-05	**0.0000e+00**	7.1986e-01
Max	9.6102e-07	6.1166e-02	1.3443e+00	1.0990e-01	1.7226e-02	1.0134e+00
Mean	6.8691e-08	9.9888e-03	1.1963e+00	8.7693e-03	2.7920e-03	8.6439e-01
SD	1.8258e-07	1.6495e-02	6.3204e-02	2.1022e-02	5.0817e-03	7.8978e-02

Table 7. Pairwise Wilcox Test - 1000 iterations

	Rand-to-Best/02/01	Rand-to-Best/02/02	Best/01	Best/02	Rand/01
		Rosenbrock			
Rand-to-Best/02	2.5367e-16	-	-	-	-
Best/01	3.7174e-11	2.5367e-16	-	-	-
Best/02	2.5367e-16	2.5367e-16	2.5367e-16	-	-
Rand/01	2.5367e-16	3.0440e-15	2.5367e-16	**1.00**	-
Rand/02	2.5367e-16	2.5367e-16	2.5367e-16	2.5367e-16	2.5367e-16
	Rand-to-Best/01	Rand-to-Best/02	Best/01	Best/02	Rand/01
		Sphere			
Rand-to-Best/02	2.5367e-16	-	-	-	-
Best/01	2.5367e-16	2.5367e-16	-	-	-
Best/02	2.5367e-16	2.5367e-16	2.5367e-16	-	-
Rand/01	2.5367e-16	2.5367e-16	2.5367e-16	2.5367e-16	-
Rand/02	2.5367e-16	2.5367e-16	2.5367e-16	2.5367e-16	2.5367e-16
	Rand-to-Best/01	Rand-to-Best/02	Best/01	Best/02	Rand/01
		Schwefel			
Rand-to-Best/02	4.8197e-15	-	-	-	-
Best/01	3.0597e-11	1.4124e-04	-	-	-
Best/02	2.5367e-16	2.5367e-16	2.5367e-16	-	-
Rand/01	2.5367e-16	2.5367e-16	2.5367e-16	2.5367e-16	-
Rand/02	**1.00**	5.0734e-16	4.0097e-09	2.5367e-16	2.5367e-16
	Rand-to-Best/01	Rand-to-Best/02	Best/01	Best/02	Rand/01
		Rastrigin			
Rand-to-Best/02	2.5367e-16	-	-	-	-
Best/01	2.5367e-16	2.9893e-03	-	-	-
Best/02	4.8197e-15	2.5367e-16	2.5367e-16	-	-
Rand/01	2.5367e-16	6.8998e-14	2.5367e-16	2.5367e-16	-
Rand/02	2.5367e-16	6.7875e-03	**1.00**	2.5367e-16	4.9465e-14
	Rand-to-Best/01	Rand-to-Best/02	Best/01	Best/02	Rand/01
		Griewank			
Rand-to-Best/02	9.3409e-04	-	-	-	-
Best/01	2.5367e-16	2.8922e-02	-	-	-
Best/02	2.5367e-16	4.2888e-10	2.5367e-16	-	-
Rand/01	2.5367e-16	4.2888e-10	2.5367e-16	7.2623e-12	-
Rand/02	**0.8777**	**1.00**	**1.00**	2.6807e-10	2.6807e-10

Table 8 shows the outcomes using 2000 iterations. We can see that the following strategies got the best outcomes: Best/01 in Rosenbrock, Sphere, and Girewank; Rand/01 in Schwefel; Rand-to-Best/01 in Rastrigin; and, Rand/01, Best/01, and Rand-to-Best/01 in Griewank. These results are similar to previous ones, excepting the /DE/Rand/01 in Schwefel function. Thus, in order to compare the results, we present the performed Wilcox test in Table 9, in which the following pairs of strategies are similar: Rand-to-Best/02 - /DE/Best/01 and /DE/Rand-to-Best/01 - /DE/Rand/02 in Schwefel; /DE/Rand-to-Best/02 - /DE/Best/01, DE/Rand-to-Best/02 - /DE/Rand/01, and /DE/Best/01 - /DE/Rand/01 in Rastrigin; and, /DE/Rand-to-Best/02 - /DE/Best/01 and /DE/Rand-to-Best/01 - /DE/Rand/01 in Griewank. Again, the odd results in

Griewank can be due to Bonferroni correction. However, other similarities do not affect the best results.

Table 8. Benchmarks Functions with dimension equals 30 and 2000 iterations

	\multicolumn{6}{c}{Rosenbrock}					
	1	2	3	4	5	6
1	1.7466e+01	**2.4787e-08**	2.5381e+01	1.4357e+01	7.1601e+00	2.2055e+01
2	2.0193e+01	4.6521e+00	2.6650e+01	2.2205e+01	1.4356e+01	2.3454e+01
3	1.8661e+01	1.1234e+00	2.6026e+01	1.7898e+01	8.4958e+00	2.2676e+01
4	7.4202e-01	1.7202e+00	3.1780e-01	1.8577e+00	1.2497e+00	3.4817e-01
	\multicolumn{6}{c}{Sphere}					
	1	2	3	4	5	6
1	2.5491e-25	**9.4259e-83**	7.8954e-06	3.8013e-17	1.3742e-40	5.1418e-09
2	5.4018e-24	2.6395e-79	5.6174e-05	6.9350e-15	3.5416e-38	1.0337e-07
3	1.6620e-24	6.0983e-80	2.5735e-05	5.5134e-16	4.4345e-39	3.0339e-08
4	1.3500e-24	8.4986e-80	1.2211e-05	1.2471e-15	7.0430e-39	2.2764e-08
	\multicolumn{6}{c}{Schwefel}					
	1	2	3	4	5	6
1	**-1.2081e+04**	-1.0903e+04	-7.9193e+03	-1.0189e+04	-1.2064e+04	-7.6214e+03
2	-1.1030e+04	-8.0430e+03	-6.5661e+03	-8.6383e+03	-1.0615e+04	-6.5821e+03
3	-1.1653e+04	-9.6618e+03	-6.9277e+03	-9.5628e+03	-1.1384e+04	-7.0611e+03
4	3.1695e+02	6.2779e+02	2.8969e+02	4.1354e+02	3.7435e+02	2.7613e+02
	\multicolumn{6}{c}{Rastrigin}					
	1	2	3	4	5	6
1	7.89	27.86	137.83	28.43	**7.50**	128.97
2	21.44	85.57	186.29	116.83	24.13	181.58
3	13.59	58.32	163.12	78.02	14.44	156.01
4	3.32	14.37	12.38	23.39	4.01	14.51
	\multicolumn{6}{c}{Griewank}					
	1	2	3	4	5	6
1	**0.0000e+00**	**0.0000e+00**	1.4379e-02	7.6272e-14	**0.0000e+00**	2.5974e-05
2	0.0000e+00	7.1154e-02	6.1266e-01	2.4573e-02	1.7236e-02	2.4379e-01
3	0.0000e+00	9.5107e-03	2.3921e-01	5.1725e-03	1.7247e-03	3.5162e-02
4	0.0000e+00	1.4312e-02	1.9017e-01	6.7582e-03	4.2493e-03	6.2286e-02

4.4. Results for EDP - 40 Generators

Table 10 shows the results for 500 iterations and the solution on each generator to attend the demand in the Economic Dispatch Problem. In this experiment, the /DE/Rand-to-Best/01 strategy presented the minimum production cost, even producing 34 KW more than the demand. Using 1000 iterations, as shown in

Table 9. Pairwise Wilcox Test - 2000 iterations

	Rand-to-Best/01	Rand-to-Best/02	Best/01	Best/02	Rand/01
Rosenbrock					
Rand-to-Best/02	2.5367e-16	-	-	-	-
Best/01	2.5367e-16	2.5367e-16	-	-	-
Best/02	2.5367e-16	7.2623e-12	2.5367e-16	-	-
Rand/01	2.5367e-16	2.5367e-16	2.5367e-16	7.6750e-06	-
Rand/02	2.5367e-16	2.5367e-16	2.5367e-16	7.2623e-12	2.5367e-16
Sphere					
Rand-to-Best/02	2.5367e-16	-	-	-	-
Best/01	2.5367e-16	2.5367e-16	-	-	-
Best/02	2.5367e-16	2.5367e-16	2.5367e-16	-	-
Rand/01	2.5367e-16	2.5367e-16	2.5367e-16	2.5367e-16	-
Rand/02	2.5367e-16	2.5367e-16	2.5367e-16	2.5367e-16	2.5367e-16
Schwefel					
Rand-to-Best/02	4.5128e-10	-	-	-	-
Best/01	4.5128e-10	**0.0714**	-	-	-
Best/02	4.5128e-10	2.5367e-16	2.5367e-16	-	-
Rand/01	4.5128e-10	2.5367e-16	2.5367e-16	2.5367e-16	-
Rand/02	**1.00**	1.7757e-15	2.5367e-16	2.5367e-16	2.5367e-16
Rastrigin					
Rand-to-Best.02.3	2.5367e-16	-	-	-	-
Best/01	2.5367e-16	**1.00**	-	-	-
Best/02	2.5367e-16	2.5367e-16	2.5367e-16	-	-
Rand/01	1.0147e-15	**1.00**	**1.00**	2.5367e-16	-
Rand/02	4.5080e-03	4.5270e-10	4.5270e-10	4.5270e-10	6.1120e-10
Griewank					
Rand-to-Best/02.4	1.5665e-06	-	-	-	-
Best/01	1.8177e-11	**0.3232**	-	-	-
Best/02	3.0440e-15	8.6955e-11	1.8177e-11	-	-
Rand/01	2.5367e-16	7.8058e-11	1.8177e-11	2.5367e-16	-
Rand/02	**1.00**	9.5428e-03	3.3075e-05	2.3652e-09	3.5812e-10

Table 11, the best result was presented by /DE/Rand-to-Best/02. Finally, using 2000 iterations as illustrated by Table 12, the minimum cost was obtained by /DE/Rand/01 strategy, even generating 17 KW more than the demand.

The Wilcox test, presented in Table 9 shows that /DE/Rand-to-Best/01 is different only compared against /DE/Best/01 in 500 iterations. In 1000 iterations, the difference is between /DE/Rand-Best/02 and /DE/Best/01. Finally, using 2000 iterations, the difference is between /DE/Rand/01 and /DE/Best/01

Table 10. EDP using 500 iterations

#Gen	1	2	3	4	5	6
1	189.6808	36.0000	114.0000	82.1994	114.0000	36.0000
2	36.0000	114.0000	152.7074	294.7477	153.4867	114.0000
3	60.0000	60.0000	60.0000	60.0000	60.0000	60.0000
4	80.0000	80.0000	190.0000	80.0000	80.0000	80.0000
5	97.0000	47.0000	49.2615	47.0000	47.0000	170.6661
6	68.0000	183.2000	68.0000	213.6797	68.0000	68.0000
7	186.0000	308.9997	110.0000	369.0970	286.2548	483.4241
8	289.3509	292.9829	300.0000	300.0000	282.9593	135.0000
9	300.0000	300.0000	300.0000	300.0000	300.0000	300.0000
10	130.0000	126.6804	130.0000	130.0000	130.0000	89.2000
11	94.0000	94.0000	94.0000	94.0000	94.0000	94.0000
12	94.0000	94.0000	94.0000	94.0000	94.0000	94.0000
13	115.2800	125.0000	50.0000	118.6848	125.0000	94.3115
14	125.0000	125.0000	125.0000	140.7174	125.0000	125.0000
15	125.0000	215.8676	125.0000	125.0000	125.0000	125.0000
16	125.0000	125.0000	125.0000	125.0000	125.0000	125.0000
17	500.0000	500.0000	500.0000	500.0000	500.0000	500.0000
18	500.0000	500.0000	493.1731	500.0000	500.0000	500.0000
19	550.0000	550.0000	550.0000	412.6551	550.0000	427.4992
20	550.0000	550.0000	550.0000	550.0000	550.0000	550.0000
21	550.0000	550.0000	550.0000	550.0000	550.0000	550.0000
22	550.0000	550.0000	550.0000	550.0000	550.0000	550.0000
23	550.0000	550.0000	550.0000	550.0000	550.0000	550.0000
24	550.0000	550.0000	550.0000	550.0000	550.0000	550.0000
25	550.0000	550.0000	550.0000	550.0000	550.0000	550.0000
26	550.0000	550.0000	550.0000	550.0000	550.0000	550.0000
27	10.0000	10.0000	10.0000	10.0000	10.0000	10.0000
28	10.0000	10.0000	10.0000	10.0000	10.0000	10.0000
29	10.0000	10.0000	10.0000	10.0000	10.0000	10.0000
30	47.0000	97.0000	169.9003	47.0000	127.9725	97.0000
31	268.0000	481.6268	520.0722	544.3848	407.9008	472.0067
32	547.2207	501.2684	60.0000	190.0000	468.4165	523.4070
33	190.0000	147.5536	190.0000	190.0000	190.0000	190.0000
34	200.0000	200.0000	485.9699	90.0000	156.0000	200.0000
35	361.4831	90.0000	487.6068	535.1799	409.6733	471.3666
36	533.3241	446.0223	153.8121	200.0000	408.7320	413.9170
37	25.0000	25.0000	110.0000	110.0000	25.0000	25.0000
38	25.0000	179.3740	236.5651	151.0442	25.0000	110.0000
39	209.2941	25.0000	25.0000	25.0000	126.3200	25.0000
40	550.0000	550.0000	550.0000	550.0000	550.0000	548.4766
Min	120370.5663	120870.2812	120795.2999	120885.0834	**120203.3804**	120479.0203
Max	123072.7495	125399.6581	122625.9715	124208.7739	123793.6618	122839.6285
Mean	121733.2207	123441.3224	121651.2147	122081.2124	121744.5554	121494.3146
SD	591.8956	1245.0497	435.6020	723.7694	890.8208	525.1857
Power	10500.6337	10500.5756	10499.0684	10499.3901	10534.7158	10577.2748

Table 11. EDP using 1000 iterations

#Gen	1	2	3	4	5	6
1	146.8954	114.0000	114.0000	146.1705	114.0000	114.0000
2	114.0000	187.9813	36.0000	36.0000	114.0000	124.6797
3	60.0000	60.0000	60.0000	60.0000	60.0000	60.0000
4	80.0000	80.0000	80.0000	80.0000	80.0000	80.0000
5	127.4672	47.0000	128.6622	47.0000	47.0000	97.0000
6	68.0000	68.0000	68.0000	68.0000	78.3680	68.0000
7	178.4000	286.5741	345.6000	186.0000	300.0000	413.2176
8	300.0000	300.0000	349.2944	510.4751	289.3805	290.6368
9	300.0000	300.0000	300.0000	300.0000	300.0000	300.0000
10	130.0000	130.0000	92.7966	87.3392	130.0000	130.0000
11	94.0000	94.0000	94.0000	94.0000	94.0000	94.0000
12	94.0000	94.0000	94.0000	94.0000	94.0000	94.0000
13	125.0000	125.0000	125.0000	125.0000	125.0000	125.0000
14	125.0000	125.0000	125.0000	125.0000	125.0000	125.0000
15	125.0000	125.0000	125.0000	125.0000	125.0000	125.0000
16	125.0000	122.0222	125.0000	125.0000	125.0000	125.0000
17	500.0000	500.0000	500.0000	500.0000	500.0000	500.0000
18	500.0000	500.0000	500.0000	500.0000	500.0000	500.0000
19	550.0000	426.8000	550.0000	550.0000	475.7795	550.0000
20	550.0000	550.0000	550.0000	493.0231	518.0668	550.0000
21	550.0000	550.0000	550.0000	550.0000	550.0000	550.0000
22	550.0000	550.0000	550.0000	550.0000	550.0000	550.0000
23	550.0000	414.4459	550.0000	550.0000	550.0000	550.0000
24	550.0000	550.0000	524.9352	550.0000	550.0000	550.0000
25	550.0000	550.0000	550.0000	550.0000	550.0000	550.0000
26	550.0000	550.0000	550.0000	550.0000	550.0000	550.0000
27	10.0000	10.0000	10.0000	10.0000	10.0000	10.0000
28	10.0000	10.0000	10.0000	10.0000	10.0000	10.0000
29	10.0000	10.0000	10.0000	10.0000	10.0000	10.0000
30	97.0000	214.5832	97.0000	151.2092	125.1489	88.9609
31	474.4982	518.7968	408.1892	493.2908	351.8803	444.9281
32	520.2775	190.0000	507.3512	375.8228	538.0512	548.8977
33	158.7086	190.0000	159.2127	190.0000	143.2000	190.0000
34	200.0000	491.6524	200.0000	200.0000	90.0000	90.0000
35	237.7083	528.8221	491.8471	499.3089	492.6522	380.2925
36	430.9853	156.0000	389.4996	536.6192	318.8000	307.7735
37	25.0000	25.0000	25.0000	25.0000	173.6618	127.0000
38	165.9668	185.4864	25.0000	25.0000	25.0000	25.0000
39	25.0000	25.0000	25.0000	25.0000	165.6481	25.0000
40	550.0000	550.0000	550.0000	408.1229	550.0000	479.5728
Min	119954.0618	120055.4902	120072.6874	119811.9265	119843.3796	**119707.8393**
Max	122279.3398	125514.0399	122063.1584	122618.2401	123452.8314	122271.9506
Mean	120896.5072	122950.4441	120984.7877	121214.5161	121070.7978	120743.0228
SD	578.0308	1259.6686	415.3477	650.9780	866.3901	533.4618
Power	10507.9073	10505.1644	10545.3883	10511.3816	10498.6373	10502.9597

Table 12. EDP using 2000 iterations

#Gen	1	2	3	4	5	6
1	36.0000	188.8800	36.0000	114.0000	160.8000	108.1992
2	114.0000	221.5892	36.0000	123.3379	102.4338	36.0000
3	60.0000	60.0000	60.0000	60.0000	60.0000	60.0000
4	80.0000	131.8977	80.0000	80.0000	80.0000	80.0000
5	125.8000	91.7634	47.0000	47.0000	47.0000	117.9806
6	68.0000	68.0000	68.0000	68.0000	116.0606	133.3184
7	267.7853	333.7155	269.9585	312.9550	336.7143	184.8847
8	300.0000	300.0000	301.2806	359.0078	135.0000	285.7440
9	300.0000	300.0000	300.0000	300.0000	135.0000	300.0000
10	68.2659	130.2682	130.0000	130.0000	130.0000	130.0000
11	94.0000	94.0000	94.0000	94.0000	94.0000	94.0000
12	94.0000	100.0440	94.0000	94.0000	94.0000	94.0000
13	125.0000	125.0000	125.0000	125.0000	123.9200	121.7370
14	125.0000	125.0000	125.0000	125.0000	125.0000	125.0000
15	125.0000	125.0000	125.0000	125.0000	125.0000	125.0000
16	125.0000	125.0000	125.0000	125.0000	125.0000	125.0000
17	500.0000	500.0000	500.0000	500.0000	500.0000	500.0000
18	500.0000	500.0000	500.0000	500.0000	500.0000	500.0000
19	426.9962	550.0000	550.0000	315.9200	550.0000	550.0000
20	550.0000	550.0000	550.0000	550.0000	550.0000	550.0000
21	550.0000	550.0000	550.0000	550.0000	550.0000	550.0000
22	550.0000	550.0000	550.0000	550.0000	550.0000	550.0000
23	550.0000	550.0000	550.0000	550.0000	550.0000	550.0000
24	550.0000	550.0000	550.0000	550.0000	550.0000	550.0000
25	550.0000	550.0000	550.0000	550.0000	550.0000	550.0000
26	550.0000	550.0000	550.0000	550.0000	550.0000	550.0000
27	10.0000	10.0000	10.0000	10.0000	10.0000	10.0000
28	10.0000	10.0000	10.0000	10.0000	10.0000	10.0000
29	10.0000	10.0000	10.0000	10.0000	10.0000	10.0000
30	47.0000	164.7731	47.0000	134.8770	47.0000	97.0000
31	452.4313	549.7642	473.3003	478.6452	483.3952	455.7047
32	460.1313	158.8000	507.1103	397.2784	543.1289	523.2376
33	190.0000	187.0348	190.0000	190.0000	190.0000	190.0000
34	200.0000	390.2687	90.0000	200.0000	90.0000	90.0000
35	422.2662	90.0000	409.5985	460.0324	476.3552	396.9631
36	539.7073	200.0000	541.6739	448.3684	387.1192	543.2072
37	106.6000	110.0000	110.0000	110.0000	176.1776	25.0000
38	110.0000	127.0861	110.0000	25.0000	110.0000	120.1248
39	25.0000	25.0000	25.0000	25.0000	25.0000	25.0000
40	550.0000	550.0000	550.0000	550.0000	550.0000	515.6652
Min	**119380.7056**	120383.8456	119881.6271	119687.9715	119850.7757	119527.2737
Max	121639.0861	124480.7678	121620.8897	123710.4819	122734.8857	121783.1923
Mean	120285.3759	122435.7239	120322.0864	121223.3933	120817.4500	120263.6570
SD	455.5893	1064.2152	324.7160	955.5607	788.7428	478.4579
Power	10517.9835	10502.8848	10499.9223	10497.4221	10498.1048	10532.7664

and /DE/Best/02. Probably these results come from the fact that all strategies can eventually produce the best result as we increase the number of iterations.

Table 13. Pairwise Wilcox Test for Economic Dispatch: 500, 1000, and 2000 iterations

	Rand-to-Best/01	Rand-to-Best/02	Best/01	Best/02	Rand/01
		500 iterations			
Rand-to-Best/02	1.00	-	-	-	-
Best/01	2.0915e-06	2.4492e-08	-	-	-
Best/02	0.557	9.1733e-03	5.2593e-04	-	-
Rand/01	1.00	0.886	2.5708e-06	1.00	-
Rand/02	1.00	1.00	2.5910e-07	0.196	1.00
		1000 iterations			
	Rand-to-Best/01	Rand-to-Best/02	Best/01	Best/02	Rand/01
Rand-to-Best/02	1.00	-	-	-	-
Best/01	4.5969e-07	1.9386e-10	-	-	-
Best/02	1.00	0.0585	9.9557e-08	-	-
Rand/01	1.00	1.00	1.3952e-09	0.7183	-
Rand/02	1.00	0.132	1.3624e-10	1.00	1.00
		2000 iterations			
	Rand-to-Best/01	Rand-to-Best/02	Best/01	Best/02	Rand/01
Rand-to-Best/02	0.0316	-	-	-	-
Best/01	7.1806e-07	8.8936e-13	-	-	-
Best/02	0.980	2.1470e-05	1.9446e-04	-	-
Rand/01	0.1798	1.00	6.8846e-13	1.4876e-05	-
Rand/02	0.669	1.00	5.2941e-13	1.2345e-05	1.00

CONCLUSION

This work presented how differential evolution tackle continuous search space. Five benchmark functions and a real-world problem named Economic Dispatch Problem were solved using 500, 1000, and 2000 iterations. As expected, solutions are getting better, and the differences between strategies tend to diminish as we increase the number of iterations. This behavior might not be the same as we increase the complexity of the problem being solved.

Regarding the quality of solutions obtained by each strategy, /DE/Best/01 and /DE/Rand-to-Best/01 tended to present the best quality solutions in benchmark functions. On the other hand, the best results were presented by /DE/Rand-to-Best/01, /DE/Rand-to-Best/02, and /DE/Rand/01 in solving the economic dispatch problem. This variety of results can be a strong indicator that changing the mutation strategy in execution-time strategy can lead to good quality solutions regardless of the problem being optimized.

REFERENCES

[1] Abbaspour K.C., Schulin R., and van Genuchten M.Th., Estimating unsaturated soil hydraulic parameters using ant colony optimization. *Advances in Water Resources*, 24(8):827 – 841, 2001.

[2] Storn Rainer and Price Kenneth, Differential evolution: A simple and efficient adaptive scheme for global optimization over continuous spaces. *Journal of Global Optimization*, 23, 01 1995.

[3] Michalewicz Z., *Genetic Algorithms + Data Structures = Evolution Programs (3rd Ed.)*. Springer-Verlag, Berlin, Heidelberg, 1996.

[4] Qing A., *Fundamentals of Differential Evolution*, pages 41–60. 2009.

[5] torn R. Sand Price K., Differential evolution a simple and efficient heuristic for global optimization over continuous spaces. *Journal of Global Optimization*, 11(4):341–359, 1997.

[6] Chen S., Montgomery J., and Boluf-Rhler A., Measuring the curse of dimensionality and its effects on particle swarm optimization and differential evolution. *Applied Intelligence*, 42, 04 2015.

[7] Jamil M. and Yang X-S., A literature survey of benchmark functions for global optimization problems. *CoRR*, abs/1308.4008, 2013.

[8] al Rifaie Mohammad Majid and Aber Ahmed, *Dispersive Flies Optimisation and Medical Imaging*, pages 183–203. Springer International Publishing, Cham, 2016.

[9] Bosman Peter A. N. and Thierens Dirk, *Numerical Optimization with Real-Valued Estimation-of-Distribution Algorithms*, pages 91–120. Springer Berlin Heidelberg, Berlin, Heidelberg, 2006.

[10] Marques I. F. C. and Cortes O. A. C., Um algoritmo hbrido baseado em fireworks e evoluo diferencial para otimizao numrica. In *Eleventh Computer on the Beah*, pages 42–44, 2020.

[11] Cortes O. A. C. and da Silva J. C., A local search algorithm based on clonal selection and genetic mutation for global optimization. In *Eleventh Brazilian Symposium on Neural Networks*, pages 241–246, 2010.

[12] Pereira-Neto A., Unsihuay C., and Saavedra O. R., Efficient evolutionary strategy optimisation procedure to solve the nonconvex economic dispatch problem with generator constraints. *IEEE Proceedings - Generation, Transmission and Distribution*, 152(5):653–660, 2005.

[13] Barros R. S., Cortes O. A. C., Lopes R. F., and da Silva J. C., A hybrid algorithm for solving the economic dispatch problem. In *2013 BRICS Congress on Computational Intelligence and 11th Brazilian Congress on Computational Intelligence*, pages 617–621, 2013.

[14] Ribeiro Jr. E. C., Cortes O. A. C., and Saavedra O. R., A parallel mix self-adaptive genetic algorithm for solving the dynamic economic dispatch problem. In *Simpsio Brasileiro de Sistemas de Energia*, pages 1–6. SBA, 2020.

[15] Corder G. W. and Foreman D. I., *Nonparametric Statistics for Non-Statisticians: A Step-by-Step Approach*. Wiley, New Jersey, USA, 2009.

INDEX

A

algorithm, v, vii, ix, 1, 3, 4, 5, 6, 7, 10, 23, 24, 25, 26, 30, 31, 33, 34, 35, 37, 38, 40, 41, 42, 44, 45, 50, 51, 52, 53, 55, 56, 59, 60, 61, 63, 65, 66, 67, 68, 69, 72, 75, 76, 77, 79, 82, 83, 87, 98
alternative hypothesis, 84
arithmetic, 53, 68
attractor, viii, 1, 3, 4, 10, 11, 21, 22, 25, 26, 27, 28, 29, 30, 31, 32, 34, 35, 36, 37, 39, 43, 44
attractor-based search system (ABSS), viii, 2, 25, 26, 27, 28, 29, 30, 32, 33, 34, 35, 37, 39, 40

B

base, ix, 46, 49, 50, 52, 53
benchmark, ix, 75, 77, 78, 79, 84, 85, 87, 96, 97
benchmarks, 78, 85
biometric, v, viii, ix, 45, 46, 48, 49, 52, 61, 63, 66, 69, 70
biometrical data, 46, 47, 49, 52
biotechnology, 77
branching, 28, 35, 36, 38, 39, 41, 42, 44

C

centrality assumption, 8
combinatorial optimization, vii, 1, 2, 3, 23, 30, 40
comparative analysis, ix, 46, 49, 61, 66, 69
complexity, viii, ix, 2, 3, 4, 29, 34, 35, 38, 39, 40, 42, 43, 61, 65, 69, 75, 78, 96
computation, vii, 1, 23, 48, 49, 52, 61, 77
computational complexity, viii, 2, 29, 34, 39
computational mathematics, 3, 40
computer, 3, 15, 26, 37, 40, 52
computing, 3, 4, 7, 24, 33, 38, 40, 53
configuration, 6, 7, 9, 14, 15, 16, 17, 18, 20, 21, 22, 25, 27, 28, 29, 30, 38, 39, 52, 77
constraints, vii, ix, 75, 80, 81, 83, 84, 85, 98
content-based image retrieval (CBIR), vii, viii, 45, 46, 47, 48, 49, 50, 51, 53, 60, 66, 73
convergence, 9, 16, 21, 22, 23, 32, 35, 77
convergence in solution, 23
convergence in value, 23
convexity assumption, 8
correlation, 10, 19, 67

D

data structure, 5, 29, 38, 41
database, 47, 50, 51, 52, 53, 54, 61, 66, 70, 72
depth, 26, 27, 28, 32, 38
detection, 49, 69
dimensionality, 78, 97
distribution, 6, 9, 16, 20, 29, 30, 59, 68, 80, 84
diversity, 29, 77
dual-tree complex wavelet transform (DT CWT), 46, 49, 51, 52, 53, 69, 72, 73
dynamical properties, 21
dynamical systems, vii, 1, 2, 10, 11, 12, 41, 42, 44

E

economics, 77
edge configuration, 6, 7, 9, 14, 15, 16, 17, 18, 19, 20, 21, 22, 25, 27, 28, 29, 30, 32, 38, 39
effective branching factor, 36, 38, 39
encoding, 48
engineering, 2, 3, 40, 47, 52, 77
Euclidean distance, ix, 46, 49, 50, 52, 53, 66, 68
evidence, 32, 35, 87
evolution, v, vii, ix, 9, 11, 20, 21, 75, 77, 79, 81, 82, 83, 84, 85, 87, 89, 91, 93, 95, 96, 97
execution, 68, 97
extraction, 46, 47, 48, 49, 66

G

generators, ix, 75, 77, 80, 81, 85, 86, 91
global feature vectors, 46

global optimization, viii, 2, 4, 5, 23, 24, 25, 29, 33, 42, 44, 82, 97, 98
globally optimal solution, viii, 2, 3, 7, 17, 23, 25, 34
graph, 3, 4, 5, 26

H

hardware, 52
Hausdorff distance, ix, 46, 49, 50, 52, 53, 61, 66, 67, 68, 69
heuristic local search, vii, 1, 2, 3, 4, 6, 7, 8, 10, 11, 41
hybrid, 26, 98

I

identification, vii, viii, 45, 47, 48, 50
illumination, 50, 52
image(s), viii, ix, 15, 16, 45, 46, 49, 50, 51, 52, 53, 54, 55, 56, 57, 58, 59, 60, 61, 62, 63, 64, 65, 66, 68, 69
imaging systems, 47
influence function, 18
intensity values, 56, 63
iris, 48, 49, 52, 56, 59, 60, 61, 63, 65, 69
issues, 3, 41, 44, 70

L

linear programming, 80
local feature vectors, 46, 49, 50, 51, 69
local search, vii, 1, 3, 4, 5, 7, 8, 9, 10, 11, 12, 13, 14, 15, 16, 17, 20, 21, 22, 25, 28, 29, 30, 31, 32, 34, 35, 36, 37, 39, 40, 42, 43, 98
locally optimal solution, viii, 2, 3, 24

M

mathematics, 30
matrix, 4, 5, 6, 9, 13, 14, 15, 16, 17, 18, 19, 20, 21, 22, 25, 27, 28, 29, 32, 34, 38, 39, 41, 72
multimodal optimization, 2, 12, 13
mutation, ix, 75, 77, 82, 97, 98

O

operating system, 10, 52
optimization, vii, 1, 2, 3, 4, 5, 12, 13, 23, 24, 25, 29, 30, 33, 34, 40, 42, 44, 76, 77, 78, 79, 80, 81, 82, 97, 98
optimization method, 80

P

parallel, 24, 26, 98
pattern recognition, 18, 61
probability, 9, 13, 14, 17, 24, 29, 84
problem space, 76
programming, 10, 80, 84
programming languages, 10
protection, vii, viii, 45
pruning, 28, 41, 44

Q

quadratic programming, 80
query, 46, 50, 53, 54, 55, 56, 57, 59, 60, 61, 63, 65, 68

S

science, 3, 40

search space, vii, ix, 3, 22, 28, 31, 34, 35, 37, 39, 41, 42, 75, 76, 77, 78, 79, 81, 84, 85, 96
search trajectory, 7, 8, 9, 10, 11, 14, 21, 22, 30, 31, 32, 36
security, vii, viii, 45, 48
solution, viii, ix, 1, 2, 3, 4, 5, 6, 7, 8, 10, 11, 12, 13, 16, 17, 18, 19, 20, 21, 22, 23, 24, 25, 26, 27, 28, 29, 30, 31, 32, 33, 34, 35, 37, 38, 39, 40, 43, 44, 48, 52, 66, 76, 80, 82, 83, 84, 91
solution attractor, viii, 1, 2, 3, 4, 11, 12, 13, 16, 17, 20, 21, 22, 25, 26, 27, 28, 29, 31, 32, 33, 34, 35, 36, 37, 38, 39, 41, 43
solution space, viii, 1, 3, 5, 6, 7, 8, 10, 11, 12, 17, 18, 19, 22, 23, 24, 26, 27, 28, 29, 30, 31, 34, 35, 37, 40
space, v, vii, viii, ix, 1, 3, 5, 6, 7, 8, 10, 11, 12, 17, 18, 19, 22, 23, 24, 26, 27, 28, 29, 30, 31, 34, 35, 37, 39, 40, 42, 50, 52, 56, 75, 76, 77, 78, 79, 81, 83, 84, 85, 87, 89, 91, 93, 95, 96, 97
structure, 5, 9, 22, 35, 42, 48, 56, 59, 65, 77

T

techniques, viii, ix, 23, 40, 45, 46, 48, 49, 79
test data, ix, 46, 50, 52, 53
texture, 48, 50, 55, 56, 61, 63, 65, 66, 69
trajectory, 7, 8, 9, 10, 11, 14, 21, 22, 30, 31, 32, 36, 43
traveling salesman problem, v, vii, 1, 2, 4, 41, 42, 43, 44

W

wavelet, 46, 51
wavelet coefficients, 46
wavelet transform, 46